Monitored

Monitored

Business and Surveillance in a Time of Big Data

Peter Bloom

First published 2019 by Pluto Press
345 Archway Road, London N6 5AA

www.plutobooks.com

British Library Cataloguing in Publication Data
A catalogue record for this book is available from the British Library

ISBN 978 0 7453 3863 7 Hardback
ISBN 978 0 7453 3862 0 Paperback
ISBN 978 1 7868 0392 4 PDF eBook
ISBN 978 1 7868 0394 8 Kindle eBook
ISBN 978 1 7868 0393 1 EPUB eBook

This book is printed on paper suitable for recycling and made from fully managed and sustained forest sources. Logging, pulping and manufacturing processes are expected to conform to the environmental standards of the country of origin.

Typeset by Stanford DTP Services, Northampton, England

Simultaneously printed in the United Kingdom and United States of America

Contents

Acknowledgements vi
Preface: Completely Monitored vii

1. Monitored Subjects, Unaccountable Capitalism 1
2. The Growing Threat of Digital Control 27
3. Surveilling Ourselves 51
4. Smart Realities 86
5. Digital Salvation 112
6. Planning Your Life at the End of History 138
7. Totalitarianism 4.0 162
8. The Revolution Will Not Be Monitored 186

Notes 203
Index 245

Acknowledgements

This is dedicated to everyone in the DPO – thank you for letting me be your temporary Big Brother and for the opportunity to change the world together.

Preface
Completely Monitored

In 2017 Netflix released the hi-tech thriller *The Circle* with a star-studded cast including Tom Hanks, Emma Watson, and John Boyega. Beneath its standard plot lies a chilling vision of a coming dystopian tomorrow. It presents nothing less than the rise of a new virulent form of tyranny where big data and social media can track anyone, anywhere, at any time. This frightening scenario may sound far-fetched but it in fact mirrors real-life developments. As reported in the *Guardian*, former Facebook president Sean Parker warned that its platform 'literally changes your relationship with society, with each other … God only knows what it's doing to our children's brains'. And while *The Circle* had a predictable Hollywood happy ending, our own future is far less assured.

Rapidly emerging is the growing threat of 'totalitarianism 4.0', one that is rising alongside the present hi-tech revolutions of 'Industry 4.0' fuelled by advances in big data, artificial intelligence, and digital communications. Rather than the ominous visage of Big Brother in *1984*, this new attempt at total control will come in the form of wearable technology, depersonalised algorithms, and digitalised audit trails. Everyone will be fully analysed and accounted for. Their every action monitored, their every preference known, their entire life calculated and made predictable. Yet this also raises a key question – who is behind this updated totalitarianism? Perhaps it is more accurate to ask who or what is benefitting from this totally monitored society? And just as importantly who and what is not being monitored and why?

The key to answering these questions is to critically explore and reconsider our common understandings of the term accounting itself. Accounting is conventionally associated with financial accounting, a fact that is not surprising given that finance has largely driven the twenty-first-century economy. However, it also refers to the collection and analysis of information about people – specifically the use of techniques to account for our beliefs and actions. Thus just as financial tools can be used to quantify and interpret the profits of a business, so to can social accounting techniques be employed to map the behaviour of people through the accumulation of their personal and shared data.

It is absolutely crucial, therefore, to better understand how the proliferation of these new accounting techniques – particularly linked to big data, social media, and artificial intelligence – are transforming the ways people are socially controlled and how, in turn, the present status quo is being reinforced. On the one hand, new technology has made it easier to track all aspects of our existence – from work to home and everything in-between. On the other hand, political and economic elites appear to conduct their business in secret, with little public oversight or knowledge. Further, the actual movement of capital and the spread of its power seems to happen in relative darkness, hidden by esoteric financial modelling and complicated accounting strategies whose primary purpose is evasion rather than detection. Significantly, in the present period financial and social accounting have increasingly merged – as the ability to collect and analyse people's data is aimed at and judged according to the same fiscal values of maximising their economic value. The overriding purpose of this book is thus to demonstrate how these accounting techniques are making the majority of people in the world more accounted for and ultimately accountable, while rendering elites and the capitalist system they profit from dramatically less so.

Being Complete Monitored

One of the most interesting and worrying features of the modern world is the ease in which personal information is obtained and exchanged. Everything from your favourite type of music to your present need for a new hammer to even your New Year's resolutions are digitally monitored and increasingly exploited by corporations and governments. Our thoughts and our actions are becoming progressively archived, as data from our past are being used to openly and not so openly shape our present and future choices. More precisely, the question is: to what extent has being made more accounted for also made us and society generally more politically and ethically accountable?

One thing is abundantly clear: it is certainly simpler to follow and judge the lives of others. It is now possible to monitor almost everything we do, from what time we wake up in the morning, to how many steps we take throughout the day, to the types of movies we binge watch at night, to the number of times we check our emails at work, to the amount of time we spend working from home.

And this information is not merely personal – it is increasingly shared for the entire world to see and analyse for their own voyeuristic and profitable purposes. Who hasn't looked up an old friend or partner on Facebook? Who hasn't Google searched themselves or those they know to discover in seconds a previously unknown accomplishment or possibly even hidden salacious secrets? And information that is private is seemingly easily uncovered by those with the technological know-how and criminal desire to do so.

At the turn of the new millennium it would appear that everyone and everywhere is, for better or for worse, more visible. This form of total personal and collective exposure has given birth to a new type of citizen. While conventional ideals of free speech, civic engagement, and social responsibility certainly have not disappeared (at least in principle), they are being enhanced and to some extent replaced by updated forms of digital morality for guiding individual and social behaviour. In particular, people are expected to properly manage their information so that they do not use it in ways that are destructive either to themselves or others. This could mean something as obvious as not posting offensive views on your social media account, or something as fundamental as regularly monitoring your heart rate. However, there is also a dark side to this digitalised citizenship. It is increasingly used to pressure people into being more productive, efficient and marketable – thus progressively making them more fiscally accounted for in their everyday actions and habits.

Underlying all these changes is the rise of a brave new world of accountability. The fact that we have so much information about ourselves and our communities means that we have no excuse not to act in a way that is not personally and economically valuable – either to yourself or your employers. There is no longer any reason to be fat given that you can count your calories on your mobile phone, and look up the nutritional content of everything you eat with the push of a button. There is no justification for being unemployed when you can create a LinkedIn account, update your CV online for prospective employers to view and build up your marketability through taking online courses. How can you possibly not get all you need done in the day when all you have to do is download a helpful 'to do' app on your phone that will practically manage your affairs for you to maximise your productivity?

Obviously these sentiments are slightly exaggerated. Still, they point to the growing relationship between being fully accounted for and being made fully accountable. Failure is attributed to one's own lack of willpower or unwillingness to gather the information necessary for your success. Equally significant, we must constantly monitor what we say and do, for you never know what from your past will come back to haunt your present. If *The Circle* threatened us with the prospect of being made 'fully transparent' – of having everything you do and say available made public – we are in danger in real life of becoming completely monitored and made 'fully monitored and accountable'.

Systematic Oversight

The hi-tech risk of total accountability is definitely real. Yet ironically it also masks a modern-day threat that is just as troubling – the power in being almost completely unaccounted for and unaccountable. While the vast majority of people across the world are directly or indirectly subjected to enhanced data collection and increased responsibility based on this information, a privileged few are escaping any such detection. The headlines are full of reports that the 1 per cent are secretly moving their money offshore to avoid paying taxes. The spread of capitalism to every corner of the world is obfuscated by esoteric financial language and models that even top graduates have trouble deciphering. If it is true that globalisation has made the world smaller, it has also rendered it much less transparent in quite profound ways.

In this spirit, there are renewed questions of what these new technologies are actually accounting for and to what social ends. What is the purpose of being more productive and does it benefit you or your employer? What are the psychological effects of these increasing demands to constantly monitor your physical health? How does this place the responsibility on you to be better while giving a 'get out of jail free card' – often quite literally – to the system and the elites who most profit from it?

Particularly, it seems that those at the top are free from such daily and invasive forms of digital scrutiny. CEOs are rarely asked how much they have worked each day or if they are being productive. US presidents can apparently spend their work time on Twitter or golfing without fear of being fired. The popular image of elites under siege by the media may have some cachet, but it ignores how little we know about their actions and intentions. It is why WikiLeaks and other types of 'open-source' subversions, while certainly ethically questionable, remain so relevant and arguably necessary. You may not like their methods, but it is undoubtedly in the public interest to know if a presidential candidate is supporting right-wing coups against foreign democracies or secretly spying on their citizens.

There is also a marked difference in how these elites are monitored and held accountable, if at all. It is now a familiar lament that those responsible for the financial crisis were not only completely unaccounted for but also not held to account for their criminal actions. It would seem that nearly causing a complete global financial meltdown was not worthy of a single trader going to jail, or that politicians who initiate costly military invasions based on false pretences never have to face a day in court.

This personal unaccountability brings to light an even more fundamental systematic oversight: capitalism itself becomes immune to any ethical or social responsibility for the international destruction it wreaks. Whether it is to our environment or the mass of the world's population, the free market is insulated from having to account for itself morally. Rather, it is shielded from such judgements by persistent claims that 'There is No Alternative'. Thus, at the beginning of the new millennium we are confronted with a strange reality in which the majority of people are called upon to be fully monitored and accountable, while the free market system and those political and economic elites who most profit from it are allowed to become ever more powerful with little to no accountability whatsoever.

Monitored Subjects, Unaccountable Capitalism?

This book explores a central contradiction of twenty-first-century economics and society: the more morally and politically unaccountable capitalism and capitalists are, the more monitored and accountable the mass majority of its subjects must become. The technocratic ideology and surveillance-heavy culture of our modern marketised societies hides a deeper reality of a free market that is unmanageable, and a corporate elite whose actions cannot be traced let alone regulated. This work aims, therefore, to highlight the paradoxical way an often disjointed and unjustifiable modern neoliberalism persists through subjecting individuals and communities to a wide range of technical and ethical 'accounting' measures, such as ever more comprehensive performance reviews and the growing use of big data in all areas of contemporary life. These pervasive and increasingly constant practices of monitoring and codifying everything and everyone mask how, at its heart, this system and its elites remain socially uncontrollable and ethically out of control.

Crucially, it provides a fresh and urgent perspective on the evolution of twenty-first-century power and resistance. It highlights the rise of 'accounting power', whereby accounting techniques are progressively deployed so that an individual's every action is measured and judged in real time in accordance with neoliberal demands for greater efficiency, productivity and profitability. The contemporary threat of totalitarianism is therefore found in the growing ability to render people 'fully transparent' and hence controllable. The new era of capitalist discipline is the ability to hold subjects internally and externally accountable, giving them a pernicious sense of fleeting control, in the face of a seemingly unaccountable and out of control global capitalism.

If this present reality seems bleak, then it also points the way to a new radical agenda for progressive change. It opens the space for challenging this paradoxical and exploitive 'accounting power' and consequently the virulent strain of neoliberalism it represents. It can inspire the channelling of technology and accounting for a social liberation that emphasises the creation of more responsive and accountable forms of administration, which support subjects who are unaccountable to capitalism and therefore more free to pursue the full scope of their personal and collective potential.

A key, perhaps defining, challenge of our time, then, is the need to overcome the creation of responsible subjects and unaccountable capitalism. Doing so means dramatically reversing who and what we hold to account and as such hold accountable. Specifically, rather than promote disciplined digital citizens – forced to exploit their personal data to maximise their economic value – it is instead critical to demand that the systems administering our lives become responsive and oriented to allowing us to explore new identities and ways of being in the world; to push for new technologies to be not just 'smarter' but more personally and socially empowering; and to require that big data and analytics hold those in power and the entrenched order responsible for their misdoings while helping to produce new, emancipated post-capitalist societies. It is nothing less than a revolutionary call for the creation of accountable systems and liberated subjects.

1
Monitored Subjects, Unaccountable Capitalism

On 8 November 2016, millions of US citizens from across the nation went to vote in perhaps the most important election of their lifetimes. Little did they know the country had already been invaded. It was not by bombs or troops. It was not an economically crippling blockade or an apocalyptic chemical attack. Rather it was a new type of weapon, one whose historical roots combined the most insidious aspects of twentieth-century covert operations with the most dangerous viral techniques of the twenty-first-century information age. In the middle of the night and in broad daylight, a secretive force had infiltrated the last remaining global superpower and had turned its citizen's data against them.

The full facts of this attack are only now coming to light. The data analytics firm Cambridge Analytica digitally harvested over 50 million Facebook profiles in order to individually target US voters for political gain.[1] Specifically, the 'CEO' of Donald Trump's campaign used his prominent position at the company to 'wage a culture war on America using military strategies' employing according to a former employee 'the sorts of aggressive messaging tactics usually reserved for geopolitical conflicts to move the US electorate further to the right'.[2] Suddenly, what seemed like harmless clicks indicating what one 'liked' were weaponised and made into a 'lucrative political tool'.[3] Indeed, these 'smart' strategies were especially effective against a formidable political machine like the Clinton and the Democratic establishment. The Trump campaign

> had bet the house on running a data-led campaign, figuring that was their best chance against the formidable Clinton machine. Cambridge were the data guys brought in to help him do it. Their main job was to build what they called 'universes' of voters, grouping people into categories, like American moms worried about childcare who hadn't voted before.[4]

Of course, the danger of Cambridge Analytica and these types of cyber-invasions goes far beyond one single election. They threaten to undermine the very survival of modern democracy itself. Already, similar methods by the same company have been blamed for swaying the shocking Brexit vote by the UK to leave the EU. 'There are three strands to this story. How the foundations of an authoritarian surveillance state are being laid in the US' quoting one popular UK commentator, 'How British democracy was subverted through a covert, far-reaching plan of coordination enabled by a US billionaire. And how we are in the midst of a massive land grab for power by billionaires via our data. Data which is being silently amassed, harvested and stored. Whoever owns this data owns the future.'[5] This new hi-tech battlefront was populated by nefarious computerised secret agents like former 'Etonian-smoothie' and big time adman Nigel Oakes, who was infamously hailed as Trump's 'weapon of mass persuasion' and the '007 of big data'.[6]

However, digging beneath the hype is an even more worrying truth. These attacks were only the tip of the iceberg as 'this type of campaign could only be successful because established institutions – especially the mainstream media and political-party organizations – had already lost most of their power, both in the United States and around the world'.[7] More than simply a loss of trust, they uncovered a brave new world where big data was 'hacking the citizenry' to shape popular beliefs and concretely reinforce existing inequalities.[8] It represented a growing form of 'evil media' able to digitally mould how people think and act, a social media virus engineered to 'manipulate the things or people with which they come into contact' for purposes of power and greed.[9] Not surprisingly, perhaps, this 'evil' was directly related to the growth of data-based academic research funded by state security agencies and the military.[10] Moreover, the reach of this surveillance was almost unprecedented – with the potential to monitor upwards of two billion people.[11]

This is a modern-day horror story where truth has become stranger and dramatically more troubling than fiction. It is full of scandal, outrage and liberal pieties about the need to protect our individual rights and sacred democratic institutions. And yet amid the noise, anger and inspiring protests, it is easy to miss the deeper reality of what is happening. Before Cambridge Analytica, before Trump and Brexit, big data was viewed as the hero not the villain. Those same voices disdaining these corrupting digital methods were once its greatest champions. As leading critical theorist William Davies recently declared:

> There is at least one certainty where Cambridge Analytica is concerned. If forty thousand people scattered across Michigan, Wisconsin and Pennsylvania had changed their minds about Donald Trump before 8 November 2016, and cast their votes instead for Hillary Clinton, this small London-based political consultancy would not now be the subject of breathless headlines and Downing Street statements. Cambridge Analytica could have harvested, breached, brainwashed and honey-trapped to their evil hearts' content, but if Clinton had won, it wouldn't be a story.[12]

It was the key to creating a sleek, efficient and bright 'smart' future. And it was by no means confined to mere elections or political campaigning. It was and is being used to reconfigure education policy – to data mine our children's personalities and emotions with the desire to predict 'national productivity in a global education race'.[13]

This reveals the ideological beating heart of big data. It is as much a promise, a technological 'myth', as it is a reality.[14] A vision is emerging of a different society where data rules our lives for better and worse. This vision can be found in the creation of 'data frontiers' for industries, portraying big data as a force for exploring and exploiting innovative ways of manufacturing not only goods but, quite literally and figuratively, the world.[15] Such changes are reflected in hopeful investments in smart technology and analytics to radically improve our lives and society. However, this promise is far from ideologically or politically neutral. Contained within its romanticised ideals revolving around speed, efficiency and innovation is an agenda that too often serves the few at the expense of the many.[16]

Nevertheless, there is a perhaps much more profound question that must be asked. What is not monitored and for what reason? It is all too common to lament that big data is just a symptom of a society where everyone is under surveillance all the time, where everything we do and think is being watched by the all-seeing eye of the digital corporate and government Big Brother. What these legitimate fears ignore though is how much of sociality remains hidden from view. From tax evasion to elite back-door deals to destroy our environment, big data has made the public little wiser about the actual people and methods used to rule our world and control our existences. Going even deeper, commonly missed among the white noise of social media, wearable technologies and the glamour of Silicon Valley is the massive amount of physical and digital labour that is being exploited to support these technologies and hi-tech cultures. It is easily forgotten, in this respect, that

> the wealth of Facebook's owners and the profits of the company are grounded in the exploitation of users' labour that is unpaid and part of a collective global ICT worker. Digital labour is alienated from itself, the instruments and objects of labour and the products of labour. It is exploited,

> although exploitation does not tend to feel like exploitation because digital labour is play labour that hides the reality of exploitation behind the fun of connecting with and meeting other users.[17]

Arguably even more terrifyingly, most of us rarely even know which data has been taken from us and to what profitable ends.[18]

The question of who and what is monitored is perhaps the defining questions of our time. In his recent book, *Master or Slave? The Fight for the Soul of Our Information Civilisation*, scholar Shoshana Zuboff warns that we are at a critical juncture:

> we have a choice, the power to decide what kind of world we want to live in. We can choose whether to allow the power of technology to enrich the few and impoverish the many, or harness it for the wider distribution of capitalism's social and economic benefits. What we decide over the next decade will shape the rest of the twenty-first century.[19]

This is undoubtedly true. But there are equally important questions that must also be asked. Notably, how does the increasing ways in which the majority of the world's population is being monitored actually contribute to an unmonitored power elite? How does this constant surveillance of our thoughts, actions and preferences lead to a capitalist system which is by and large left unsurveilled? How is this culture of monitoring progressively colonising and exploiting not only current realities but our virtual ones as well? And finally, how have we been socially produced to become ultimately our own personal customisable twenty-first-century 'Big Brothers'?

Aim

This book aims to theoretically and empirically reimage capitalism by offering a novel perspective on the development of modern power as it attempts to control a progressively data-based and virtual population. It critically investigates the paradoxical relationship between personal accountability and systematic unaccountability in contemporary neoliberalism. It reveals that ironically, as capitalism becomes less accountable in terms of its practices and values, individuals within this system become increasingly monitored and made accountable regarding their beliefs and practices. In this respect, sophisticated financial accounting techniques have made capitalist transactions more esoteric, and given elites greater opportunities to hide their profits through techniques such as tax avoidance and evasion. Significantly, this has played into a prevailing belief that despite its clear and present problems, capitalism cannot be altered and is therefore largely morally unaccountable for its destructive economic, social and political effects. Simultaneously, the rise of big data and social media have rendered the majority of individuals more accounted for in terms of how they spend their time as well as their daily behaviour. This has, in turn, forced them to be more accountable (both to themselves and those in authority).

At stake is the evolution of power and control for a digital world. Rather than being confined to the physical environment, market domination extends into our virtual realities. Capitalism is no longer satisfied with simply exploiting our labour – it now wants to shape and proscribe the limits of our multiple selves in cyberspace and beyond. It is coding and profiting from our diverse datafied identities and is pre-emptively colonising any computerised or simulative world we can conceive of. And ironically, it is relying on us more than ever to accomplish this total economic and social conquest. We are its data explorers – dispatched to discover new virtual markets and 'smart' data-driven profitable opportunities. And we are the ones who must constantly monitor ourselves and these multiple realities to ensure that they conform to these overriding fiscal prerogatives. In this new age of big data, you can increasingly imagine anything you like and be anyone you want, just so long as it expands the bottom line.

Monitoring Society?

It seems clear that in the present era we are being watched and analysed more than ever. While previous periods certainly desired knowledge about the world and the people who inhabited it, for both cultural and technological reasons they paled in comparison to the contemporary drive to be 'totally informed'. At its most pure, it follows an Enlightenment tradition to clarify our given reality, to bring light to areas of understanding that remain dark. Moreover, it seeks to use data to reveal previously unseen aspects of our individual and human condition. Amid the numbers are clues and patterns that can alter how we see each other and our very existence. Yet it also raises the question of who is in control of this information, who is driving its collection, and for what reason. As even the famously technologically friendly former US President Barack Obama warned, 'The technological trajectory, however, is clear: more and more data will be generated about individuals and will persist under the control of others.'[20]

This growing worry points to the complete colonisation of our lives by surveillance. The so-called big data revolution is constantly expanding, desiring to know ever more about who we are and what we will be. The inspiration for these questions is almost entirely market driven – associated with the overriding aim to maximise productivity, efficiency and profitability. To this end, 'there are now very few significant interludes of human existence (with the colossal exception of sleep) that have not been penetrated and taken over as work time, consumption time, or marketing time'.[21] These ultimately narrow objectives further reveal just how much is missed by an overreliance on big data. In the efforts to obtain limitless information the richer context is easily and often overlooked, as are alternative forms of knowledge that could challenge these hegemonic market blinders.[22]

This mass infusion of data into traditional market ideas and practices has been presciently described as 'surveillance capitalism'. Personal information is now a prime resource to exploit and commodify. As such the rise of big data signifies 'a deeply intentional and highly consequential new logic of accumulation that I call surveillance capitalism. This new form of information capitalism aims to predict and modify human behaviour as a means to produce revenue and market control.'[23] Consequently, humans become the creator, product and consumer all at once. We produce our own data, we are produced as datafied goods and we ravenously buy back this information about ourselves. Thus the new capitalist behemoths like Facebook 'are part of a heavily personalised, data-intensive economy that exploits the digital labour of its user base'.[24]

Central to this digital exploitation is simply how enjoyable it can feel and ultimately addicting it can become. We are constantly clicking, refreshing and checking up on our datafied selves. The mobile phone is now so prevalent it is close to being a permanently visible appendage for people. There is always another clickbait article to read, more information to discover, steps to count, movie reviews to critique and restaurant locations to find. And with each digital encounter we are being technologically exploited more and more. These often hidden economic demands on ourselves certainly take their mental and physical toll. Internet addiction and overuse is now a certifiable condition that requires social prevention and medical treatment.[25]

Why then do so many of us continue to do it? What lies in our individual and collective compulsion to be ever more connected and updated? To understand this conundrum, it is essential to grasp the ironically empowering aspects of this domination. American writer Bruce Schneider speaks thus of a 'hidden battle to collect your data and control your world', and 'that in half a century people will look at the data practices of today the same way we now view archaic business practices like tenant farming, child labor, and company stores'.[26] Still, it is a 'bargain' we presently make based on widespread desires for the convenience it provides from corporations and the protection it offers from governments. The attractiveness of big data and its personal use therefore extends far beyond the horizon of a future digital utopia. Rather, its enjoyment is experienced in the here and now, as 'Self-tracking has to be understood in relation to behavior that is predominantly about getting things done in ways that are possible, suitable and meaningful for the individual.'[27]

What is absolutely key is that our surveillance is never complete. It is always both partial and perennially unfinished. There will never be a moment in which CEOs and politicians, and even radical hackers, stop and say 'we have collected enough data – our job here is done'. Instead it is ongoing and exponential. Each new dataset, each fresh piece of information, each novel algorithm is simply the means to collecting and analysing more. And there is a fundamental human element to this smart culture – namely, we are ultimately responsible for its continual and constant collection. While much of this data gathering is hidden and automatic, it relies on people to not only provide such raw material but find innovative ways for its expansion. This is reflected in an emerging form of 'surveillent individualism', according to scholar Shiv Ganesh, 'which emphasizes the increasingly pivotal role that individuals play in surveillance and countersurveillance, [and] is central to understanding the ambiguities and contradictions of contemporary surveillance management'.[28] Consequently, we are increasingly becoming not so much 'quantified selves' but, more accurately, 'quantifying selves'.[29]

Appearing before us is a culture revolving around regular, systematic and ever larger monitoring. It is at once exploitive and empowering, ever-present and increasingly unintrusive. Yet as we enter this monitored society, it is unclear whether elites or the system itself is becoming more accounted for or accountable. Further, this surveillance era, for all its information, seems to have made our everyday realities less rather than more clear. Ironically, as we fragment into increasingly small data-byte selves and identities, the oppressive system and power differentials driving this process are solidifying, unmonitored, behind the scenes.

Monitoring (Post)Modernity

Conventional understandings of domination focus almost exclusively on the shaping and controlling of a person's identity and actions. It presumes, even if only implicitly, a coherent self – as prevailing ideologies and status quos mould people into their powerful images. Yet the digital age challenges this traditional perspective. This is the era of intersectionality, of multiple selves, of pluralism in who one is and strives to be. We are expected to increasingly 'have it all', to resist being confined to any one identity. This reflects, in part, how post-modern ideas have gone mainstream. The twentieth-century notion of a 'unified' self is being rapidly replaced. The present age is witnessing

> the reformulation of the self as a site constituted and fragmented, at least partially, by the intersections of various categories of domination/oppression such as race, gender, and sexual orientation. Thus, far from being a unitary and static phenomenon untainted by experience, one's core identity is made up of the various discourses and structures that shape society and one's experience within it.[30]

While there are obviously many reasons for this shift, the intervention of technology is clearly prominent among them. In particular, the growing presence of data, virtuality, computers and robotics is evolving previously sacred natural assumptions regarding the body and personhood. Put differently, we are no longer seen as being simply organic. Rather, at play is the 'Reconfiguration of the body as [the] combination of "technological" and "biological"' both increasingly in fact and in the popular imagination making it 'not as a fixed part of nature, but as a boundary concept'.[31] The philosophical railing against essentialism is being realised to a large extent by technological advancements that render selfhood artificial and therefore both changeable and plural.

One specifically arising phenomenon is that of rebooted 'digital selves'. With the power of social media it is now possible to inhabit many identities at once. It is an avatar culture, where sophisticated games and digital communication has allowed us to take on a range of different identities.[32] The popular game *Second Life* provides a revealing glimpse into this rapidly emerging world of digital selves. Here, people can choose a brand new life by selecting a fresh identity and playing it out online in real time. More than just escapism, anthropologists describe it as a modern form of 'techne', denoting the 'the bootstrapping ability of humans to craft themselves'.[33] Nevertheless, these created selves are still influenced by a person's social context and biases. Recent research found, for instance, that 'although Second Life provides unprecedented freedom in appearance, local social contexts, as much as external ones, created powerful boundaries and expectations, leading many participants to seek [a] socially acceptable appearance that would be interpreted in certain ways as part of their interactions'.[34]

These selves are therefore connected but not coherent. They are diverse expressions of a common living dataset. Consequently, the established notion of the free agent must be reconsidered if not entirely rebooted. The new generation is composed of digital selves navigating a vast and expanding cyberspace. Our common humanity is not as thinking and acting rational decision-makers but as multiple users surfing the web. 'The self is increasingly digitised in a number of identities, accounts or profiles related to engagement with social, public and commercial services', according to Carlton et al. 'These identities are multiplied across the civic, social, commercial, professional and personal contexts of their use, and the vulnerabilities of this atomised citizen are not well understood'.[35]

At the core of these selves is deep-seated insecurity. Identity, of course, is rooted in a sense of inadequacy, the desire for belonging, and a longing to discover 'who one really is'. The sociologist Erving Goffman's famous 'dramaturgical analysis' spoke to such needs long before the advent of the digital age.[36] He observes how we craft front and backstage selves – for public and internal consumption. Updated to the present, our digital masks hide and seek to cope with growing feelings of personal fragmentation and subjective incoherence.[37] These anxieties are only exacerbated by these data-driven transformations to our daily existence, anxieties particularly acute and common during times of rapid technological change.[38] The era of identities as avatars and profiles produces as much disquiet as it does excitement.[39]

Importantly, our digital selves are being progressively enhanced by our emerging virtual realities. We become socialised as adept citizens of these digitally mediated cultures. It is a hi-tech existence marked by processes of online attachment, splitting and self-concealment.[40] More and more we embrace the fact that 'we are data', as our offline selves disappear, a relic of an earlier unconnected time.[41] There are no clear front and backstages, just digital platforms upon which we can make ourselves more and less visible. Illuminated instead is a 'transmedia paradigm' that stands 'as a model for interpreting self-identity in the liminal space between the virtual and the real, [which] reveals a transmediated self constituted as a browsable story-world that is integrated, dispersed, episodic, and interactive'.[42]

Hidden in this 'smart' life of concealing and exposing oneself are the unmonitored forces guiding our preferences and practices. Corporations have developed sophisticated techniques to take advantage of our digital selves. This includes using specially created, customisable, 'virtual selves' to influence your physical behaviour and buying choices.[43] In this respect, we are entering into unregulated digital spaces with often invisible perils and unseen forces of exploitation and manipulation.[44] In guiding ourselves through this largely unregulated cyberspace we easily miss just how socially constructed these selves still remain.

At stake is the transition of modern capitalist domination to a brave new post-modern digital world. Selfhood is now plural, online selves for us 'smartly' to control. We can creatively play with identity, creating second, third and fourth lives. We are cyber-personas, which can express a broad array of human emotions and subject positions, from supportive Facebook friend to villainous anonymous website troll. What unites all of these identities is their shared ability to be externally monitored and profitably exploited.

Accounting for Neoliberalism

The contemporary period, for all its diversity and unpredictability, is primarily marked by neoliberalism. Notably, across the globe and within different contexts and cultural histories, there is a drive for greater marketisation, privatisation and financialisation. The crash of 2008 and the Great Recession which followed it has perhaps slowed down these trends, as well as given them a strong ideological challenge, but they have by no means deterred them or put an end to them. Its spread relies upon not simply overt governing structures but also the creation of a 'governable' market subject linked to everyday practices of power.[45] As such it is a 'mobile technology' that exists 'not as a fixed set of attributes with predetermined outcomes, but as a logic of governing that migrates and is selectively taken up in diverse political contexts'.[46]

Yet it is precisely this mobility that also links neoliberalism so inexorably to big data. We live in an increasingly mobile society, where smartphones are ubiquitous and digital communication alters the very ways we engage with each other as humans. Capitalism has therefore had to construct a social technology that matches and can take advantage of this dynamic mobile technology. Already we are witnessing the collapse of the public and private sphere, in which the public becomes privately owned and our private lives become a matter of public scrutiny.[47] Social media has made it possible for employers and governments to 'know you', often better than you know yourself, and to use this information for their own gain. It is what leading surveillance theorist Kirstie Ball refers to as an 'all consuming surveillance' that matches consumer preferences with corporate interests.[48]

This is reflected in the total exposure of these digital selves to market-based desires and judgements. It seems we are entering into the 'age of digital transparency', where 'our digital selves will have personalities that are accessible to anyone who cares to look. These will be more revealing than a conversation with us, and more accurate than our own hopes and desires'.[49] More than just being technologically vulnerable, these traces produce the very material in which we are accounted for and made accountable to an all-pervasive neoliberal rationality of profit and productivity. Returning to the movie *The Circle* highlighted in the Preface, this dark satire of Silicon Valley, based on the book by noted author David Eggers, reveals 'the ease with which we relinquish our freedom, and our lives, to corporate control'.[50]

Humankind has returned in a sense to its past nomadic ways. We easily traverse across vastness of cyberspace, 'shrinking' the actual world through fostering instantaneous digital connections that transcend geopolitical borders. The internet is our passport to explore different cultures, perspectives and interests. Our digital selves are gateways into speaking with multiple voices and from various points of view.[51] Nevertheless, this fluidity is undermined by the constant digital traces we leave behind.

These 'footprints' have to be constantly managed, both with regard to what is online and who can see it.[52] Those undergoing dramatic life transitions, such as transgender individuals, show in acute detail the crucial need to monitor your past and present profiles.[53] The digital surveillance of one's selves is now a near universal feature of the post-modern techno age.

In a sense, this represents a profound evolution to the 'post-human'. What currently matters are our data trails and digital tracking. If capitalism is legitimately critiqued for being dehumanising, turning us into efficient profit-making machines, then capitalism 2.0 will be remembered for being datafying. We are vessels of continuously refreshing information that can be data mined for ever greater material gain. It is our very diversity and uniqueness that makes us so valuable, as our individuality is commodified into a customisable data product that contributes to the wider, only partially visible, global e-marketplace. In the words of the brilliant theorist Rosa Braidotti:

> Advanced capitalism is a difference engine in that it promotes the marketing of pluralistic differences and the commodification of the existence, the culture, the discourses of 'others', for the purpose of consumerism. As a consequence, the global system of the post-industrial world produces scattered and poly-centred, profit-oriented power relations.[54]

Worse, we become enraptured by this mobile neoliberalism – ready nomadic travellers along its circuited, electronic, data-driven highways. We even craft our lives to meet these demands, using our data to track out the marketability and exploitability of our intersecting digital histories. Our destination is no longer an exotic trading locale but the desperately pursued but never fully reached states of maximisation and optimisation. In accounting for our lives we become always and forever accountable market subjects.

Contradictory Data

The creation of a 'datafied' society is often viewed as being wholly novel. It is a brand new reality for a hi-tech smart age. However, historically data have always played a part in the constitution of society and its power relations. The strategic deployment of information for purposes of domination is by no means unique to modern times – though it has massively advanced. This book will argue that we are now living in 'monitoring' times, where capitalism and the inequalities it relies upon are reinforced through the constant monitoring and innovative exploitation of our expanding data selves and virtual realities. Yet to fully grasp this era it is critical to situate it within the broader development of social power, particularly as it relates to the development of a tension-filled market order.

Since almost its inception, capitalism has been wracked by tensions and contradictions. While it grew out of the ashes of the philosophical Enlightenment and its political revolutions, it seemed to serve primarily the emerging bourgeois ruling class. It spoke of shared progress, but was marked by previously unheralded forms of industrial deprivation. These concrete incongruities between rhetoric and reality revealed the fundamentally conflictual character of capitalism, and the centrality of these contradictions for driving its survival and growth.

The most central and famous of these contradictions is the one associated with class. Marx, in particular, foretold of the eventual and inevitable collapse of capitalism due to its internal class contradictions. This prediction went beyond mere denunciations of worker exploitation. Instead it declared that the insatiable profit drive of the capitalist class would inevitably lead to mass unemployment and in time full-scale proletariat revolt.[55] These theories have, in turn, been undermined to an extent by the failure of capitalism to yet fall, linked its social resilience and adaptability to changing cultural, political and economic conditions.[56]

It is easy perhaps to retrospectively scoff at the failure of this Communist revolution to occur, or uncritically praise the resilience of capitalism. Yet doing so misses the important role of class struggle for shaping our market societies, both historically and looking to the future. Indeed, the present age is still marked by popular anger at 'capitalist oligarchs' and their complicit political handmaidens. The election of CEO presidents only reveal the constant ideological innovation needed to sustain this tension-plagued free market system.[57] The evolution of capitalism is one of finding continual justifications for privilege and exploitation – ranging from social Darwinism, colonialism, white supremacy and laissez-faire economics in the nineteenth century, to meritocracy, globalisation, systematic racism and monetarism in the twentieth century, to personal responsibility, smart development, white male privilege and neoliberalism in the twenty-first century.

At its heart, capitalism is defined by crises politics. Just as there are market cycles of boom and bust, so too are there cycles of capitalist legitimacy. Each new attempt at capitalist legitimisation follows a circular path of acceptance and challenge. It is matched by a progression from optimism to pessimism, as it relates to a fresh market fantasy of progress that gradually and ultimately always turns into a living social nightmare. Open Marxists, in particular, have highlighted this formative political dimension of capitalism – noting the morphing of organic crises linked to economic downturns into politicised upheavals that can be co-opted by capitalists for the system's renewal.[58] Consequently, 'such a "political reading" of crisis theory eschews reading Marx as philosophy, political economy or simply as a critique. It insists on reading it from a working-class perspective and as a strategic weapon within the class struggle'.[59]

What follows, then, is a capitalism that has both an eternal foundation in inequality and oppression and yet must forever remake itself to meet inevitable social resistance against its dominance. It would thus be misleading to suggest that capitalism is unalterable – that its cycle of crises politics simply represents an eternal return of the same. Instead, each period of rise and fall – every attempt to justify capitalism anew – reflects both changing cultural and technological conditions as well as novel political and civic constraints to its power. To give one example, the liberal consensus of the immediate post-war era represented a combination of triumphalism in the face of global military devastation, the depression and the Holocaust, alongside the growth of mass media and demands for civic equality. Each iteration of capitalism is therefore a refraction of its actually existing material and a discursive condition inexorably linked to but never completely determined by what has preceded it. It is crucial then to study the attempts to this underlying contradiction of market-based privilege and exploitation.

In this spirit, the fundamental tension in capitalism is concretely manifested in a range of evolving historically specific contradictions. Its hegemony is defined by its articulation and management of these prevailing opposing forces. A classic example is the simultaneous need for a strong state in support of a private market economy.[60] Coming from a slightly different perspective, the renowned scholar Daniel Bell speaks of a pronounced cultural contradiction plaguing mature market economies – notably how 'the unbounded drive of modern capitalism undermines the moral foundations of the original Protestant ethic that ushered in capitalism itself'.[61]

What binds these together – connects them despite their historical and often rather dramatic contextual differences – is their rootedness in perpetuating privilege. Capitalism is often quite rightly critiqued for its perpetuation of economic inequality, one that extends to and is bolstered by disparities in social and political power. These have led to sustained and growing charges against the racial, gender and geographic privilege that perpetuates these unfair differences. To this end, running parallel to the class contradiction articulated by Marx is one of accounting and accountability. It is the constant struggle for deciding who is accounted for, in what ways, and as such who and what is held politically and morally accountable.

Thus, lurking alongside these more obvious forms of entrenched advantage around class, race, ethnicity and gender is similarly pervasive and insidious form of privilege: namely, the diverse impact that new technologies and discourses have for reinforcing these material and cultural power imbalances. In the nineteenth century, how did social Darwinism advantage the new bourgeois ruling class while keeping down the rising proletariat masses? The answer lies largely in the construction of innovative accounting technology linked to an ideology of 'meritocracy' and personal responsibility for one's moral and economic fate. Nevertheless, these same accounting technologies also created new capitalist-based accountability measures for the bourgeois, around their contribution to the firm's overall profitability. The advancement of national economic models further shifted this accountability to governing elites, as they were now obligated and judged against their ability to produce economic growth.[62]

These social accounting technologies and the accountability regimes they produce and perpetuate form key parts of a culture's 'imagined communities'. The famed social anthropologist Benedict Anderson described these imagined communities – associated primarily with the rise of modern-day nationalism – as the discursive creation of a collective identity around abstract concepts.[63] While Anderson stresses the romantic and positive sense of belonging provided by these imagined relations, they are also marked by a profound sense of shared justice and progress revolving around monitoring principles and techniques. It is about manufacturing individuals and groups as particular types of social subjects through accounting for and ultimately holding them accountable for their beliefs and actions. The collection of information and its analysis is hence the daily means by which this imagined identity is given physical form and materially/culturally reproduced. The struggle for dominance between classes or groups is, in this respect, an ongoing conflict of who should be monitored and for what socio-political ends.

Updated to the present, 'smart' age, big data plays a similar role to that in the past, though with a crucial new twist. The hidden algorithms that increasingly shape our lives and choices are central to the construction of our twenty-first-century imagined community. They remain largely invisible, yet constitute the basis by which we connect to others, share a sense of identity and judge them. While perhaps not as evocative as the singing of a national anthem, social media networks and mobile communities link us with people we have never met nor probably ever will. In doing so, it places us into a broader online community where we can supposedly forge our own allegiances and enemies. Moreover, we are encouraged to become active data subjects, part of a global movement of users all trying to improve themselves through these technologies. These regular processes of self-tracking are thus daily affirmations that this 'smart' community exists and that we are part of it.

This analysis raises profound critical questions of which big data only scratches the surface. In the new millennium of 'advanced capitalism', what technologies and discourses have been discovered and promoted to cover over and strengthen the market's fundamental contradictions of inequality and exploitation through holding us accountable? Equally importantly, to what extent are the capitalist tools and ideas emerging out of these contradictions being used to disempower the many for the profit of the few? What else do they reflect about our current historical situation and the potential for future liberation?

Digitally Accounting for Neoliberalism's Contradictions

It is now almost common sense to claim that we live in a 'free market' society. Indeed, if the last three decades have had an abiding theme it is the insatiable spread of capitalism to all areas of society and all corners of the world. Resistance to this seemingly inevitable march to total marketisation is viewed as either 'idealistic dreaming' or terroristic barbarism. In the wake of the Great Recession, however, fresh questions are being asked about the nature and desirability of this complete capitalist transformation. It has raised renewed concerns over how this change is impacting society, both present and future. More precisely, what is this hyper-capitalist nightmare that we have suddenly found ourselves trapped in, and how can we escape it?

As discussed, the present age reflects a distinct shift from previous capitalist periods, representing in particular the evolution from liberalism to neoliberalism. Whereas the previous era was characterised by public welfare, government intervention and strong unions, the current one promotes trickle-down economics, privatisation and employability. It represents, in this regard, 'a theory of political economic practices that proposes that human well-being can best be advanced by liberating individual entrepreneurial freedoms and skills within an institutional framework characterised by strong private property rights, free markets, and free trade'.[64] While rhetorically valorising human liberty and extolling its commitment to individual freedom, the most prominent characteristic of neoliberalism is in fact 'The corporatization, commodification, and privatization of hitherto public assets [that] have been signal features of the neoliberal project. Its primary aim has been to open up new fields for capital accumulation in domains formerly regarded [as] off-limits to the calculus of profitability.'[65] The state, in turn, is thought to have been reduced to a mere shadow of itself, confined to a basic watchman type role.

Nevertheless, its implementation, operation and legitimisation is not so smooth or simple. It is wracked by internal tensions and external challenges to its dominance. The introduction of the market to all areas of modern existence was put forward as a cure all for all of life's social ills. The trains aren't running on time – privatisation will fix that. Disappointed with public services? Contracting them out to a private company will improve everything. We were promised a more dynamic, competitive and streamlined society refashioned in the image of the free market. It is morally, politically and economically

> grounded in the 'free, possessive individual', with the state cast as tyrannical and oppressive. The welfare state, in particular, is the arch enemy of freedom. The state must never govern society, dictate to free individuals how to dispose of their private property, regulate a free-market economy or interfere with the God-given right to make profits and amass personal wealth. State-led 'social engineering' must never prevail over corporate and private interests.[66]

There was just one not so minor – in fact quite major – problem: the market didn't work nearly as well as advertised. This failure to fully launch raised a profound contradiction for neoliberalism. Namely, who was to blame for these systematic failures? Put differently, the entire discourse surrounding the free market began to revolve around questions of accountability. This dovetailed nicely with its original emphasis on the self – interest and personal responsibility.[67] These discourses provided the justification for the dismantling of the post-war welfare state, emphasising individual achievement and downplaying any sense of collective responsibility.

This focus on responsibility, of course, became even more important as the cracks in the once sacred free market began to show. The boom and bust of the 1980s gave way to seemingly unending economic growth in the 1990s. However, this prosperity was a chimera, masking rising inequality and chronic economic insecurity. There were also renewed concerns regarding the negative economic, political and environmental impact of corporate globalisation. The elite rejoinder that the international spread of the free market was 'inevitable' may have been accepted, but was hardly inspiring, especially to its growing number of victims.

Addressing these mounting issues required reasserting the primacy of personal responsibility. The need to stay competitive in the global marketplace was outsourced to individuals retraining themselves for the 'jobs of the future'. The problems of global inequality were laid at the doorstop of the 'bad governance' of poor countries.[68] Significantly,

> [f]rom the early 1990s onwards, the call for less state has gradually been substituted by a call for a better state. This new approach should not be confused with a plea for a return to the strong (Keynesian or socialist) state. Rather it implies better and transparent governance of what is left of the state after neoliberal restructuring has been implemented.[69]

Whether individual or collective, the ethos remained the same – any failures were the result of personal laziness, incompetence or malfeasance, and were therefore certainly fundamentally problematic. At the heart of the modern capitalist project was a constant shifting of blame from the shoulders of elites to those already most oppressed by the weight of systematic oppression and exploitation.

Yet it also reflected a deepening contradiction of present-day neoliberalism. The very question of responsibility, even when aimed at the most vulnerable and usually least culpable, opened up space for targeting those at the top of the political and economic pile. Indeed, even in the 1980s the corporate scandals that plagued the 'masters of the universe' were quickly followed by a fresh call for 'corporate responsibility'.[70] Broadly, it forced governments to take on new and not necessarily reduced roles, going from welfare provider to mass-market educator. In this respect, it was now the state that was responsible for teaching people the skills to be personally responsible.

This contradiction, however, was reawakened in the wake of the 2008 global financial crises and the Great Recession that followed it. Suddenly, the tables had turned and it was CEOs and financial leaders who were being asked to account for their actions. Irresponsibility became progressively associated not with the lazy welfare recipient but the neoliberal robber barons of the new gilded age. The immediate response by those in power was to, not surprisingly, either accept the need for limited reform or blame the whole problem on the 'greediness' of past governments. Nevertheless, even these reformers played into a powerful crisis narrative that married economic recovery with recovering the past optimism in the market and its ability to provide for a prosperous shared future. From more conservative and reactionary corners the fault lay with greedy individuals (especially poor ones that spent beyond their means) and profligate governments. The demand for austerity was thus as much a moral one as it was an economic solution.

At stake, therefore, was how to manage this fundamental neoliberal contradiction. Notably, how to ensure that all responsibility was directed at individuals and market enemies rather than the system and its elite profiteers. If the early period of neoliberalism was defined by theories of 'trickle-down economics', its more recent version was characterised by a chronic embrace of 'trickle-down responsibility'. This revealed a new capitalist paradox – the more that values of responsibility were touted, the less the market was held responsible for its social, economic and political costs.

Reinforcing this ironic use of accounting was an entirely new form of social technology: personal monitoring inexorably linked to big data. It followed a logic of taking personal responsibility for physical, mental and social circumstances. It allowed people and organisations to have a much fuller 'account' of one's existence and to evaluate it accordingly. However, it also rebooted the culture of accountability, situated now in the ways that people managed their digital selves. Hence, 'Within online ecosystems the real self bears special psychological-ontological characteristics where the main rule is "whoever is not available on the internet does not exist". Users mix conscious decisions with random ones, drifting along the dataflow.'[71] Accountability, as such, is increasingly connected to our diverse online 'personal brandings'[72] and ability to navigate often complex data surveillance regimes.

These practices critically bring to the fore a central contradiction of neoliberalism: who is being monitored and how are they manipulating it to their advantage? Perhaps even more fundamentally, how is the monitoring of individuals and communities used paradoxically to ensure that capitalism remains systematically unmonitored in terms of its social, political and economic effects?

The Paradox of Business and Surveillance in our Times

This book explores a key paradox linked to business and surveillance in our times – importantly, there is a direct relationship between this simultaneous increase in personal monitoring and overall systematic unaccountability. Structurally, this paradoxical dynamic serves as a means for elites to assert enhanced control over a population while concurrently freeing themselves to maximise their profit with little or no formal and informal public oversight. Psychologically, this offers individuals a greater sense of daily control over what feels like an increasingly out of control capitalist world. In doing so, it empowers people in a way that enhances their exploitation and reproduces the very system that is responsible for their oppression. Hence, the more unaccountable capitalism is allowed to be, the more accountable its subjects must be.

In doing so it critically explores of how discourses of monitoring and the concrete techniques associated with them function to ideologically reinforce and structurally reproduce a fundamentally unaccountable modern hyper-capitalist order. In this spirit, it reveals how daily demands for subjects to be more transparent, predictable and controllable in their preferences and actions ironically permits the contemporary free market and its financial and corporate beneficiaries to be less transparent, more unpredictable and largely socially uncontrollable.

It is imperative, therefore, to illuminate how the proliferation of every new accounting technique – particularly linked to financial modelling, big data and social media – are transforming how capitalism is reinforced and how the people within it are being socially controlled. These accounting techniques include sophisticated financial modelling, the introduction of algorithms to organise employment and the use of analytics driven by big data driven to shape how we work and live. What makes this book so timely is that it reveals how the deployment of these accounting techniques makes people more accountable, and the capitalist system dramatically less so.

In doing so it explicitly reveals a central tension of the modern age – how is it that individuals and communities seem to be ever more accountable, while at the broader level capitalism and capitalists are increasingly viewed and lamented as inherently unaccountable? Why, for instance, isn't the enhanced use of data collection and analysis being directed at making markets less volatile rather than simply making us more predictable consumers? Why is it allowable for corporations and governments to monitor their workforce and citizens to an ever greater extent, and yet corporate and political elites remain relatively protected from such invasions of privacy? Why is it acceptable for individuals to be constantly called to account and take personal responsibility for their actions at work and home while the global 'race to the bottom' perpetuated by international elites is viewed as unstoppable, regardless of its irresponsible and damaging environmental, political and economic consequences? Through directly addressing this contradiction and these questions, this book seeks to challenge this unaccountable capitalist system of individual and collective accountability – turning monitoring into a revolutionary tool for radical change.

2
The Growing Threat of Digital Control

Amazon is one of the richest and most popular companies in the world. It is renowned for being a pioneer in conducting digital business. Its website makes consumption as easy as a click of a button, and its use of big data helps to refine your buying preferences and deliver your items to your door, sometimes on the same day. Yet beyond the screen there lies a much darker present-day dystopia. Its workers are paid low wages to work long hours, and the same tracking technology that makes its customers' lives so easy end up making their warehouse employees miserable. They are timed to the second between stocking items and penalised if they go below the optimal speed. Every day they must pass through an intrusive security process just for the fortune of working ten and a half hours with only a 30-minute lunch break for relief. Even worse, their employees are left in a state of constant insecurity, as their jobs are 'zero hour' and rarely, if ever, permanent. Through it all they are reminded that they do not work in a warehouse but a 'fulfilment centre' where all their dreams are meant to come true, and are reminded that 'We love coming to work and miss it when we are not here!'[1] At the same time, Amazon employs its big data capabilities to 'stalk' its customers[2] and its political clout to avoid paying taxes.[3] Welcome to the gigabyte economy.

In a age when supposedly nothing is secret anymore and everything is transparent, there is much that still remains hidden from mainstream view. The 'smart economy' is characterised by precarious labour, tedious routinisation and a lack of opportunity for upward advancement. Supporting the digital services and mobile technologies that cater to all our modern needs is an army of underpaid and exploited workers toiling in the shadows.[4] These invisible men and women entered into the public consciousness when a Chinese factory making iPhones experience a rash of employee suicides, all driven to the edge by the intensified pressure of meeting consumer demands for the new update.[5] It was reported that 'Worker after worker threw themselves off the towering dorm buildings, sometimes in broad daylight, in tragic displays of desperation – and in protest at the work conditions inside.'[6] In response, all Foxconn (the company who owned the plant) did was install nets outside to catch the bodies and force all employees to sign a pledge that they would not kill themselves. According to one employee, 'It wouldn't be Foxconn without people dying. Every year people kill themselves. They take it as a normal thing'.

On the other side of the digital class divide, there is just as much that is left unseen by the masses. CEOs and corporate board members largely act in secret, with precious little government or employee oversight. Economic and political elites often form an 'inner circle' where they continually support each other for their own mutual benefit, beyond the prying eyes of the wider public.[7] Their misdeeds are left unreported unless they cause a scandal too big to ignore, at which point they are asked to apologise and accept millions in stock options and severance pay to leave quietly.[8] In 2017, 'a secret job board' previously reserved for exclusive top executives and other elites was partially opened up to the 'masses'.[9]

Yet more than perhaps anything else, it is our secrets that today's elites desire the most. They want to unlock our preferences and find the most efficient ways to profit from our likes, dislikes and everyday activities. To do so they make this largely economic endeavour into a ethical obligation and cultural necessity. It is required to be happy, healthy and personally fulfilled as well professionally successful.[10] This tracking society strategically combines the voluntary and involuntary, the seen and unseen. To this effect,

> Whether we intentionally self-track, or are tracked with or without our consent, our personal data – often of the most intimate and private nature – connects us to wider social systems. Our data contains a virtual, if partial, version of the self – a 'data double' living on servers around the world. When it travels, a part of us does, too. In this way, our data has a social life. It is both personal and political at the same time.[11]

Importantly, such tracking has been translated into daily monitoring practices where the majority of individuals are continually asked to account for themselves. In the workplace, this means accounting for how you spend your time and whether it contributes to the organisation's bottom line along with your own future marketability. Wearable technologies turn us into our own worst managers, introducing 'a heightened Taylorist influence on precarious working bodies within neoliberal workplaces'.[12] We can now assess and be judged on whether we performed a task fast enough, and more broadly whether outside of work if we are keeping healthy enough to perform our job at an optimal level. Nevertheless, there is also much that is covert about this exploitative tracking. Managers, for instance, often introduce games to instil in their workforce company goals and values. Such games may look innocuous, but they are 'rooted in surveillance; providing real-time feedback about users' actions by amassing large quantities of data and then simplifying this data into modes that easily understandable, such as progress bars, graphs and charts'. Moreover, they operate through 'The pleasures of play': 'the promise of a "game", and the desire to level up and win are used to inculcate desirable skill sets and behaviours'.[13]

This reveals the growing threat of twenty-first-century digital control. It is one where every action can be knowingly or secretly monitored.[14] It is an 'iSpy' era in which surveillance is close to omnipresent and works in obvious and covert ways.[15] The quantified self has expanded into all areas of human existence, from work,[16] to the gym,[17] to the doctor's office.[18] More than just being quantified, we have become 'datafied' – bits of information are used to regularly and continually judge our actions so that we can evolve into the perfect and whole free market subject.

A Brief Accounting of Digital Capitalist History

The popular perception of big data is that it is an unprecedented force for both social good and ill – one so unheralded that nothing in the past comes close to matching to it. Indeed, it is trumpeted by corporations and media tastemakers alike as the revolutionary missing link to future success. In 2013, then CEO of IBM Ginni Rometty declared that 'Data is becoming a new natural resource. It promises to be for the twenty-first century what steam power was for the 18th, electricity for the 19th and hydrocarbons for the 20th'.[19] Despite its pretensions of radicality, this game-changing technology sounds strikingly similar to conventional desires to use data for maximising profit. A *Washington Post* article republishing this IBM report noted that 'Businesses are grappling with how to gain better insights from the big data explosion so they can move faster and better serve their customers ... the challenge is to find fast efficient ways to glean knowledge from all that information to create a smart company.'[20] Moreover, its innovative uses are most often directed at rather traditional capital desires, such as finding oil[21] and drawing on the 'final frontier' of space data to improve businesses.[22]

It is not surprising, then, that *Wired* magazine would have a headline that read simply 'Big Data, Big Hype?'[23] Or that researchers studying something as sophisticated as 'cognitive big data' would conclude 'that the idea of Big Data is simply not new'.[24] What is novel about this data-driven economy is not so much its underlying principles of exploration and exploitation, but rather how much more infinite in scope it has the potential to become. According to the *Economist*:

> Data are to this century what oil was to the last one: a driver of growth and change. Flows of data have created new infrastructure, new businesses, new monopolies, new politics and – crucially – new economics. Digital information is unlike any previous resource; it is extracted, refined, valued, bought and sold in different ways. It changes the rules for markets and it demands new approaches from regulators. Many a battle will be fought over who should own, and benefit from, data.[25]

This idea was reinforced by the World Economic Forum, which pointedly observed: 'So the next time you hear someone say "data is the new oil", ask them when the earth will have no more data to extract and see if you get an answer'.[26]

Tellingly, what truly binds big data to historical technologies is not just its profit-making and exploitive potentials, it is also the ways it opens up fresh techniques for surveilling and regulating individuals for these overarching purposes. '(The famous Marxist) Rosa Luxemburg once observed that capitalism grows by consuming anything that isn't capitalist', writes well-known critical technology author Ben Tarnoff: 'Historically, this has often involved literal imperialism: a developed country uses force against an undeveloped one in order to extract raw materials, exploit cheap labor and create markets. With digitization, however, capitalism starts to eat reality itself. It becomes an imperialism of everyday life – it begins to consume moments.'[27]

As impressive and frightening as this sounds, it actually follows a well-worn path of deploying these technological advancements to strategically monitor individuals and populations. Big data is merely the latest and most sophisticated part of a longer story revolving around 'the rise of the information state' dating back at least until the start of the sixteenth century.[28] The creation of steam technology or the cotton gin both massively sped up production methods while creating new demands for and innovative solutions to the supervision of workers and slaves, respectively. It reflected the broader function of bosses, which is to maintain hierarchical control for the sake of maximising these productive gains.[29]

The space of capitalist production, of course, has since its very inception been a site of daily struggles over such surveillance. One of the first reasons to create a public police force in early nineteenth-century Britain was to put down resistance from workers over their conditions and their demands for greater power. These same battles would evolve later in the century into an ever-expanding system of secret police and direct supervision over 'radicals' such as Marxist and Anarchists. Less dramatically, factories required innovative 'internal control systems' to monitor their workforce, constantly having to update them in light of not just new surveillance technologies but the ability of workers to undermine these efforts.[30]

These growing efforts to monitor people and their actions, however, intensified with the rise of mass media and electronic surveillance methods. The so-called 'surveillance society' arose from the emergence of this all-seeing 'electronic eye'.[31] To a certain extent, these electrified methods reproduced traditional forms of 'bureaucratic control'.[32] Yet their difference from what had come previously is that they signified the creation of technology explicitly made for enacting and enlarging such monitoring techniques. Whereas previously technological development was aimed at enhancing production, it now encompasses the very methods by which it would ensure that such manufacturing advancements were properly put into practice by employees. Equally, this perceived need to accurately account for the actions of your workforce, combined with established surveilling cultures such as those common in prisons and other state institutions, helped create a society that was always being watched to ensure that people were being 'good citizens'.[33]

What this reveals is a central feature of what is being called 'data capitalism'. It is defined as:

> a system in which the commoditization of our data enables an asymmetric redistribution of power that is weighted toward the actors who have access and the capability to make sense of information. It is enacted through capitalism and justified by the association of networked technologies with the political and social benefits of online community, drawing upon narratives that foreground the social and political benefits of networked technologies.[34]

It is a 'revolution' that promises and threatens to 'transform how we live, work, and think'.[35] The scope of this rebooted free market society is vast and diverse. It represents all at once a 'cultural, technological, and scholarly phenomene', in that it shapes our society, our capabilities and even our broader ways of studying our world.[36] Yet it also poses a distinct problem of who controls this widespread and seemingly ubiquitous phenomenon. A key question for modern subjects is: 'am I being controlled by data or am I in control of my data'?[37]

The novelty of capitalism 2.0 is again not in its business practices or ultimate goals. Indeed, for all its sophisticated technology, 'platform capitalism' is to a certain extent little more than a repackaged 'smart' form of monopoly capitalism, where large firms can buy up smaller emerging competitors.[38] Its originality lies in its reconfiguration of the relationship between surveillance and control. At play is a new type of market-based regulation and monitoring referred to as 'dataveillance'. Here all our actions and preferences are collected, evaluated and used against us – both in overt and hidden ways – to shape our behaviour in line with elite prerogatives.[39] To this end, it can be considered a novel form of 'neuropolitical control' as it is seeking to reprogramme our brains at the neurological rather than the conscious or affective levels.[40]

In turn, digital classes are produced, separated not just by wealth but by their ability to harness and direct this smart technology for the purpose of exploitation. In the complexity of big data its rather simple core is easily ignored and intentionally obfuscated – it remains a world divided by haves and haves not. The renowned critical theorist Nick Dyer-Witheford highlights the rise of the 'cyber-proletariat', arguing that 'Class has become ontologically not less, but more real, more extended, entangled, ramified and differentiated – and yet without abolishing the opposition of exploiter and exploited on which it is posited, which is generative of countless intermediate forms, and yet preserves its simple, brutal algorithm.'[41]

Here big data is socially weaponised as a means to reinforce existing power differentials and material inequalities.[42] They create the conditions for people to actively participate in their own hi-tech class domination through being the very vessel through which the data necessary to maintain this oppressive system of control is obtained.

Surveilling Progress

Big data is quickly emerging as a tool for regulating and ultimately controlling our thoughts and actions. Drawing upon the information gained from our individual and collective daily activities, it encourages and directs us to be better buyers, workers and 'good citizens' in the market. It operates on the level of our communication, conscious decisions and non-conscious neurological processes. Such dataveillance portends not only a dystopian tomorrow but an ethically troubling and repressive today. It is also, perhaps ironically, the most potent discourse of shared progress and personal empowerment that currently exists. We are promised a 'smart world', where monitoring is put to good use to make us all richer, happier and healthier.

Consequently, it is crucial to understand the affective appeal of this pervasive data-driven surveillance. Doing so means moving beyond conventional accounts of both capitalism generally and 'data capitalism' specifically. In particular, it is imperative to interrogate how the values associated with these systems reinforce their emotional and psychological hold over the people who populate them. Take individualism, for example, which is traditionally viewed as central to the spread of the free market. Nevertheless, there is no necessary or essential connection between the 'sovereign individual' and capitalist reproduction. Rather they form a contingent social relation, where associating freedom with individuals serves to affectively legitimise a complex system of economic exploitation involving a wide range of collective bureaucratic organisations (such as firms and those linked to the state). This social rather than essential relationship is borne out in how different cultures 'customise' capitalism to their specific cultural context – exemplified by China's combining of free market principles with more communal and state-based values.[43]

Similarly, contemporary surveillance is made palatable through a diverse range of affective discourses. The most obvious, in this respect, is connecting this monitoring to popular celebrations of being watched as a means for gaining fame and notoriety. The popularity of reality TV shows reveal in stark detail how surveillance mechanisms are translated and justified via a voyeuristic culture in which people not only accept being watched by millions, but actually welcome it.[44] More practically, hidden algorithms are portrayed as being the key to providing people with better decision-making power in all areas of their lives.[45] 'Big social data' is likewise trumpeted as a 'trending' phenomenon that holds both alluring 'promises' and exciting 'challenges' for researchers and policymakers alike.[46]

These positive portrayals, of course, have done little to deter the widespread fears surrounding big data. There is an increasing, and legitimate, sense that these tech industries – once meant to 'save us' – have in actuality taken over the economy, our creativity and soon our jobs as well as our very existence.[47] And this is just the tip of the iceberg. The imagined future will be a dystopian nightmare, where we are ruled by robots and left to scrape out meek material survival in the face of mass unemployment.[48] Returning back to the here and now, it is undermining our ability to alter this seeming inevitability democratically, leading to a bigger battle between 'the people vs. Tech'.[49]

On the flip side, big data points to a possible 'smart utopia' where society will be run more efficiently for the benefit of everyone. The acclaimed writer Anthony M. Townshend preaches the gospel of 'smart cities' and the possibilities of crafting a 'new civics for a smart century', where 'putting the needs of people first isn't just a more just way to build cities. It is also a way to craft better technology, and do so faster and more frugally'.[50] While Townshend and others speak of the power of 'civic hackers', the promotion of 'smart cities' also acts as a potent form of modern-day 'corporate storytelling', where the interests of large firms are presented within a broader narrative of technology-driven shared urban progress.[51] Additionally, it signified a compelling development discourse, linking the creation of 'smart cities' in countries like India to 'new urban utopias' based on 'entrepreneurial urbanization'.[52]

Underpinning these dystopian fears and utopian desires is a hi-tech surveillance culture that enrols us into these monitoring systems of exploitation through either ignoring their effects or seducing us into their opportunities for our personal betterment.[53] The advent of machine learning inexorably linked to big data and algorithms can be harnessed by individuals to teach them how to profit from the 'digital power shift'.[54] To this end, ideas of using big data for creating a 'smarter' society is a mentality – or a 'smartmentality' – that creates policies and popular ideas which, on the one hand

> support new ways of imagining, organising and managing the city and its flows; on the other, they impress a new moral order on the city by introducing specific technical parameters in order to distinguish between the 'good' and 'bad' city. The smart city discourse may therefore be a powerful tool for the production of docile subjects and mechanisms of political legitimisation.[55]

Yet these mobile data-based technologies also encourage us to ignore their enhanced surveillance capabilities and practices.[56] Just as there is power in being able to ignore calls and texts, so too do people feel empowered by consciously using their 'smart gadgets' for their everyday enjoyment and professional success without having to consider its wider invasion of their privacy. In this way, daily items such as loyalty cards form 'local narratives' of personal ignoring where individuals voluntarily and often enthusiastically embrace consumer practices that secretly collect mass amounts of their data which are then 'traded globally without much concern by the consumers themselves'.[57]

In this spirit, practices of data monitoring have become an integral part of individual desires for agency and popular demands for progress. For instance, new apps provide parents with the ability to track their teenagers, including one ironically called 'Teensafe' that 'can monitor your child's phone without their ever knowing about it, and gives you an all access pass into all text messages (including deleted ones), their web history, their call logs and Facebook and Instagram feeds. You can also, of course, use GPS to track their every move.'[58] According to its CEO Rawdon Messenger, 'It's not about knowing who their friends are, it's purely about keeping them safe, checking that they got wherever they were going ok and knowing that they're not being bullied. This is about keeping your child safe and watching out for them.'[59]

More menacingly, this same 'spyware' technology gives 'abusers a terrifying new toolbox to control their partners and exes. Phone software allows them to follow people's movements, monitor their calls, texts and emails – and even watch them'.[60] This gives an insight into the invasive parts of a deeper 'surveillance – industrial complex',[61] where individuals and capitalism now share an unquenchable thirst for data.

Insatiable Data

A defining tension within capitalism is the relationship between its infinite desire for profit and the limited resources it has to achieve this aim. Marx referred, in this regard, to the 'insatiable' quality of the market and its elites, as their thirst for exploitation could never be fully satiated. He declares, 'Capital is dead labour, which, vampire-like, lives only by sucking living labour, and lives the more, the more labour it sucks', and moreover that this 'vampire will not let go' as the daily exploitation of workers 'only slightly quenches the vampire thirst for the living blood of labour'.[62] Yet the advent of the age of big data changes this equation, as capitalism finally confronts a resource that is just as limitless as its own desires. Significantly, this unquenchable hunger extends as much to the many as it does to the few, since unlike the past, where people may tire of work, people's need for data appears increasingly inexhaustible.

Driving this inexhaustible need for data is the so-called 'knowledge economy'. Traditionally this refers to the oversized influence of digital technologies for transforming the economy and social relations. However, it also reflects a novel ethos regarding how we see and understand the path to individual success and collective prosperity. Greater information is portrayed as the primary force for making these aspirations a reality – and as such, the more knowledge you have the better off you will be. Sensing this shift, at the beginning of the new millennium the renowned scholar Nigel Thrift predicted the growth of 'knowing capitalism'.[63] However, far from describing a market dystopia – where everything is simply a brutal economic calculus – he discusses how data and other smart technologies are used for people to 'know' more about their existence: what makes them happy, joyful and sad, and what can they predict will do so in the future. It is about plural 'knowledges', and the ways in which they intersect so as to produce both relative stability and productive tensions that can alter a given status quo. For this reason he downplays the importance of 'finance' and 'information technology' to an extent, observing instead that 'What is most interesting about contemporary capitalism is how these juggernauts of finance and information technology and regulation have interwoven with new developments to produce new possibilities for profit.'[64] These data-based potentialities, of course, also open the door to innovative forms of 'mass observation',[65] through which surveillance is redefined as a process of ongoing discovery.

In the same way that information technology produces different, and not always complementary, 'knowledges', so too does it manufacture diverse and at times divergent desires. The invisible quality of the algorithms which are progressively 'sinking in' and 'sorting' our everyday existence has fostered renewed longings for greater transparency and 'participatory web cultures', where it is humans who are ultimately in control.[66] To this end, there is a 'data revolution' occurring that seeks out the creation of more 'open data', but also wants 'better data' that can ably adopt 'hybrid approaches that mix big and small data methods'.[67] Tellingly, while this 'revolution' can certainly alter how we are governed, what we value, and how we relate to one another, it also reveals the initial efforts of the capitalist system to co-opt and find new ways to profit from values associated with collaboration and openness.

It is crucial to resist, therefore, simply equating this insatiable hunger for data with the emergence of big data technologies. Rather, it is historically connected to the neoliberalism which gave it life. This updating of traditional capitalism featured an interweaving of data technology and the desire to capitalise on every aspect of human life. Indeed, while the introduction of 'high speed computer networks' and complex modelling 'became critical mechanisms for the newly created speculative markets', this was ultimately only their most superficial social effect. Instead

> financialization's encouragement of surveillance capitalism went far deeper. Like advertising and national security, it had an insatiable need for data. Its profitable expansion relied heavily on the securitization of household mortgages; a vast extension of credit-card usage; and the growth of health insurance and pension funds, student loans, and other elements of personal finance. Every aspect of household income, spending, and credit was incorporated into massive data banks and evaluated in terms of markets and risk.[68]

From these insights, it is tempting to conclude that big data is a homogenising force, where all areas of existence must conform to pre-existing financial standards. Yet it would be more accurate to say that what this 'datafied' neoliberalism accomplishes is the mining of value from our differences and uniqueness. It customises our exploitation according to our exhibited preferences and lifestyle choices. In this sense, 'social media surveillance is a form of surveillance in which different forms of sociality and individuals' different social roles converge, so that surveillance becomes a monitoring of different activities in different social roles with the help of profiles that hold a complex networked multitude of data about humans.'[69] It even seeks to go beyond our present horizons, predicting and creating value out of our hypothetical futures.[70]

Significantly, this cultural data addiction has been progressively justified as a required part of short- and long-term social progress. Put differently, it has become a veritable 'public good' that is necessary for keeping us safe and secure from the threats of terrorists and everyday criminals.[71] Yet it has also expanded the very physical scope of public governance. The necessity of collecting data has permitted power holders to obtain this information anywhere and anytime, in extremely sophisticated ways. This expansion is exemplified in the 'politics of verticality', in which domestic drones are used to monitor populations from the air, applying hi-tech digital techniques such as 'hologrammation' that can combine multiple photographs in order to create a more accurate depiction of what is occurring 'on the ground'. This permits governments, in turn, to apply 'surgical killings' from above.[72] More broadly, the public demand for more data has created 'global assemblages' of intersecting governance promoting monitoring that transverses existing political and geographic borders.[73]

Absolutely imperative to this infinitely expansive regime of data power is the willing participation of those subjected to its rule. It is, of course, completely understood that capitalism has always enrolled the masses into ironically desiring their oppressions. The worker longs for the next promotion or finding deeper spiritual meaning in their work. The consumer seeks 'retail therapy' in their purchases, associating these economic exchanges with their short- and long-term personal happiness and well-being. In the time of big data, it is the thrill of discovering more about ourselves and the world that makes this often hidden exploitation at once so appealing and insidious. The daily enjoyment of finding out new things about our environment, what we are watching, and what we could soon do if we so choose serves to make us complicit in our own dataveillance.[74] The extension of this big data economy is grounded on our 'immaterial labour', the daily ways in which we innovatively and continuously collect data about ourselves and knowingly or unknowingly share it with corporations and governments.[75] In this respect, we are data explorers, always searching out new data frontiers for those in power to monitor and exploit.

Yet this exploration is based as much on our deep-seated and ideologically constructed insecurities as they are on the abundance of data opportunities that are now available to us. There is always a feeling that all our problems could be easily solved through more data. This extends to individuals and businesses alike. In the words of Duke Professor Dan Ariely, 'Big data is like teenage sex: everyone talks about it, nobody really knows how to do it, everyone thinks everyone else is doing it, so everyone claims they are doing it.'[76] Far from feeling disenchanted from this complex and often fragmented data-based world, we are passionately and often desperately enhanced with its digital possibilities for improving our lives and society.

The elixir of big data, smart technologies, artificial intelligence (AI) and hidden algorithms are powerful, almost magical forces that are incomprehensible yet the key to our salvation. Our longing for this technological deliverance is intimately bound with the ever-growing need to collect more data about ourselves and others. More importantly, it is the very foundation upon which capitalism's insatiable need for data and profit is transformed into an infinite contemporary demand for digital monitoring.

Monitoring the Dialectic of Digital Control

The rise of the big data society and the surveillance culture accompanying it is intricately tied up to the social conditions – specifically of neoliberalism – from which it emerged. Consequently, the insatiable desire for data did not arise in a vacuum. Our never-satisfied appetite for information stems from the very ways in which data technologies and the free market have reordered, or more precisely fractured, contemporary society. The constant aggregation of data and individuals as data has led to the general disaggregation of society. Everything and everyone now is separated into their component parts, split into their various likes, dislikes and diverse daily activities. It is precisely this culture of disaggregation that creates the (post-)modern dialectic of ever-expanding social monitoring.

Tellingly, capitalism is conventionally accused of 'rationalising' society. It views the social as a space that must be properly organised for the sake of efficiency, productivity and profit. This rationalising ethos extends beyond the workplace and encompasses issues of crime, healthcare and leisure.[77] Yet this drive towards greater order is undercut by the market's own commitment to competition, a value even more prized under neoliberalism. The responsibility of governments to support this competition and its own surrender of public oversight powers contributed, in turn, to the 'end of organized capitalism'.[78] Suddenly, what once seemed organised and stable was in flux and difficult to make any coherent sense of. These premonitions were only exacerbated by the regular economic crises and financial crashes periodically afflicting this (dis)order.

Ironically, it is exactly this perception of chaos that encourages the need for enhanced monitoring. The more disorganised a system appears, the greater the desire for it to be properly accounted for, and in doing so be made coherent and whole. It is not surprising, in this respect, that the discourses associated with social belonging are most prominent during times of social dislocation. Discourses of nationalism, ethnicity or even personalised professional identities provide individuals and communities with 'ontological security' that makes sense of and give order to their otherwise confusing reality and often unconnected experiences.[79] In practice, according to acclaimed sociologist Anthony Giddens 'the plethora of available information is reduced via routinised attitudes which exclude, or reinterpret, potentially distributing knowledge ... avoidance of dissonance forms part of the protective cocoon which helps maintain ontological security'.[80] This translates concretely into renewed monitoring regimes, put in place to continually and performatively safeguard this ontological security.

In the present context, capitalism finds itself in a rather strange predicament. The free market and the competitive ethos it promotes demands a rather large degree of 'disorganisation'. Attempts at coordination could lead to stifling the relatively free rein of corporations and the elites who run them. Further, their power is legitimised based on their being considered 'victors' in the brutal 'dog-eat-dog' world of the contemporary marketplace. The traditional organising force of the state, moreover, has been thoroughly defanged to prevent it from regulating this capitalist oligarchy. It is not surprising, then, that there has been a resurgence in ideological fundamentalism – whether attached to the market orthodoxy of neoliberalism[81] or virulent forms of religious extremism – alongside national and global crusades against existential 'terrors'.[82]

The arrival and increasing prominence of big data brings with it even greater complexity to this problem. The fragmentation of people and things into databytes makes explicit this sense of disorganisation and lack of wholeness. While data analytics makes sense of our information, it divides us into smaller and smaller components. We can now be diversely categorised, made into 'selves' rather than a 'self', as highlighted in Chapter 1. It is this profound literal and figurative disaggregation that produces mass and elite desires for enhanced data monitoring. The use of big data and digital surveillance methods to completely account for our actions grounds us in a sensible world and provides us with a sense of instrumental purpose and coherence. Data, as such, both pulls us socially apart and continually puts us back together. We track ourselves so that we do not lose ourselves.

Of course, such monitoring is never politically or ideologically value free. We have entered, according to famed scholar George Ritzer, into the next age of capitalist development called 'prosumption', where the activities of production and consumption merge into one.[83] In the digital era such prosumption has taken on new characteristics. We now are constantly both producing and consuming data, leading the majority of us to remain as both 'powerless tools of capital' and 'capitalism's creative tools' at the same time.[84] Just as one was once made 'personally responsible' for their success in the traditional economy, we are now all expected to be suitably 'self-entrepreneurial' in the digital economy.[85] Significantly, neoliberalism both before and after the emergence of big data used discourses of 'responsibilisation' as a disciplining tool in order to make people account for their actions in the face of first 'disorganised' and now 'disaggregated' capitalism.

What seems to be emerging now is an updated dialectic of capitalist control in which the more it disrupts societies and the people's lives within them, the greater the perceived need for rationalisation and monitoring. The more 'disaggregated' it becomes, the greater the need for digital surveillance and control. The individualism and self-absorption so central to contemporary neoliberalism is grounded in a monitoring culture where surveillance is used to both regulate populations and provide subjects with a sense of ontological security in a society where traditional community networks, civic relationships and public institutions are in decline. There is a renewed demand from both the top and the bottom for new and innovative accounting techniques to help stabilise this precarious sense of self. Surveillance acts, in this sense, as an often a welcome source of 'social sorting', confirming our place in a sensible and coherent social order.[86] Amid the rise of big data, information serves to re-form us as selves that we can regularly reinforce through our personalised data collection.

Surveillance is, hence, transformed into an exercise of personal exploration and self-exploitation. The so-called 'electronic panopticon' of computer screens and videos in the sky lent themselves to a different type of dialectic – one where fresh monitoring technologies and techniques had to be created simply to keep up with the various forms of popular everyday resistance to their incursion into people's privacy.[87] By contrast, digital monitoring methods now have to adapt to the innovative ways that individuals self-track and share their data publicly. These everyday data explorers present new and inviting challenges to those in power, who seek to profit off this constant flow of personalised information. The insatiable desire for data, thus, produces an equally insatiable demand for monitoring.

Virtual Power

We increasingly live in a 'monitored' world. Yet what does this actually mean for the exercise of power and control? The answer may seem to be rather obvious, as it is commonly assumed that surveillance shapes our behaviour and directs what we can and cannot do. However, the 'datafication' of society and the subjects who inhabit it has made the exercise of power much more complex than it may first appear. In particular, it is now not only productive but utterly and totally creative and adaptable. While it does try to regulate individuals, this monitoring is also about encouraging them to be different and try new things for the sake of collecting more data on them, and in doing so discovering fresh ways of exploiting them.

Traditionally, monitoring is inexorably linked to practices of coercion and discipline. Perhaps the most influential study of this phenomenon was Foucault's critical analysis of the prison panopticon, which sought 'to arrange things [so] that the surveillance is permanent in its effects'.[88] Yet even he recognised the dynamism of early surveillance regimes, presciently observing:

> If the economic take-off of the West began with the techniques that made possible the accumulation of capital, it might perhaps be said that the methods of administering the accumulation of men made possible a political take-off in relation to the traditional, ritual, costly, violent forms of

> power, which soon fell into disuse and were superseded by a subtle, calculated technology of subjection.[89]

This disciplinary society existed and evolved throughout the twentieth century, evolving and adapting to the diverse needs of bureaucratic organisations and later post-bureaucratic firms.[90]

However, the new millennium has brought with it fresh challenges and opportunities for monitoring power. In particular, digital technologies are changing the social landscape of cultural control. It is both internalising and externalising it – uniquely 'customising' it to individuals while obsessively focusing on their 'objective' data. Importantly, it has made surveillance quick, continuous and convenient. Indeed we are now increasingly part of 'surveillance assemblages' that 'works by abstracting human bodies from their territorial settings into discrete flows that are later reassembled into data doubles'.[91] As such, it is presently a progressively seamless part of one's consuming experience, turning it, consequently, into what appears to be an activity based on personal enjoyment and consent. To this end, 'Digital technologies have made it possible to govern in an advanced, liberal manner, providing a surplus of indirect mechanisms that translate the goals of political, social, and economic authorities into individual choices and commitments'.[92]

While this may be an overly optimistic account of contemporary processes of domination, it certainly speaks to a shift in emphasis from surveillance to monitoring. The former implies a close, almost obsessive tracking of a person's actions. The latter denotes a systematic accounting for their activities and conduct over time for purposes of quality assurance, and if necessary correction. These are obviously not mutually exclusive concepts but rather complementary ones. All surveillance will have an element of monitoring and vice versa. The era of big data is marked by various monitoring regimes and techniques – ones which combine a high level of regularity and systemisation with the flexibility and freedom to allow people to be their own data explorers. Critically, this rebooted monitoring culture is in many ways less concerned with what people are presently doing but rather what they probably will and could do, using their data to plan in advance how they can profit off of all their potentialities.

Nevertheless, this relative freedom should not be confused with an unregulated or non-disciplinary society. Instead, discipline is shifting and expanding into novel and interesting directions. On the one hand, the introduction of digital technologies such as computers reinforced more coercive forms of surveillance. In the new digital 'sweatshops' such as call centres 'the agents are constantly visible and the supervisor's power has indeed been "rendered perfect" – via the computer monitoring screen – and therefore making its "actual use unnecessary"'.[93] If anything, such repressive and explicit oversight of employees is only intensifying, as the example of the Amazon warehouse presented at the beginning of this chapter reveals in stark and depressing detail.

However, it is also developing new resources and goals associated with monitoring. It is becoming 'user led', following their lead in determining their preferences and desires in order to meet their data needs. In doing so, it opens up new 'personalised' data markets and sites for data-based regulation. It is vitally important that such monitoring also fosters new regimes of personal responsibility and accountability. All our actions must be optimised with the guidance of data, and further should contribute to our overall well-being – whether personally, professionally or in relation to wider society. In both instances of surveillance and monitoring our data become a continually updating benchmark on which to judge us, revealing in real time our daily progress and our failures to fully maximise our potential.

These forms of digital control are underpinned to appealing affective promises of data empowerment. Most obviously, perhaps, is the association of 'smart' technology with social, organisational and personal advancement. The failures of these technologies often to deliver on these lofty promises is attributed to human error – either at the level of the individual or existing authorities. What is crucial, in this regard, is that these smart techniques always stand on the horizon, presenting an eternally elusive goal to pursue and form our identity around. Its very disappointment is precisely what French psychoanalysts would refer to as its ironic 'jouissance' or enjoyment, as it represents our continual ontological security in this ongoing pursuit of psychic wholeness. Monitoring has thus morphed into a new cultural fantasy, representing 'The element which holds together a given community [that] cannot be reduced to the point of symbolic identification' acting in this capacity, as 'the bonds linking together its members always implies a shared relationship to the Thing, toward enjoyment incarnated … If we are asked how we can recognise the presence of this Thing, the only consistent answer is that the Thing is present in that elusive entity called our 'way of life'.[94] The constant tracking of oneself through data feeds into this monitoring fantasy, creating daily reaffirmations of the possibility of one day perfectly harmonising all of our aggregated parts into a psychic whole.

It also fosters in people a longing to be monitored, specifically to be surveilled by others. Scholars have increasingly challenged Foucault's original focus on the 'panopticon', concentrating instead on the 'synopticon', reflecting a contemporary 'situation where the many see the few to the situation where the few see the many'.[95] In the age of Facebook, YouTube and Instagram, our perceived success is intimately linked to how many people literally and figuratively watch us. With so many opportunities to watch others, and in the face of feeling our personhood disintegrate into mere databytes, knowing that people like our 'content' reinforces our specialness. When the many can now see the many, to be one of the few chosen to be given particular attention over others signifies for the world and ourselves our uniqueness.

We are entering the age, therefore, of what I refer to as 'virtual power'. It is a building upon and expansion of more discrete and physical forms of power. Yet it differs in a number of key and significant ways. First, it is often unseen and exists within the 'virtual realm' of hidden algorithms, faraway data processors, augmented reality, AI and invisible data plunderers. Second, it feeds off our potentialities as opposed to prevailing realities, monitoring all our current and possible selves and futures. Critically it also subtly and not so subtly guides these 'virtualities' to be eternally accountable to market demands of efficiency, productivity and profitability. As such, in this brave new virtual world you can in principle be anything you want, just as long as it is fiscally viable and valuable. Privilege is repackaged as to who does and does not have this digital freedom to be their own data explorers – to be monitored rather than closely surveilled. In each case, though, our social construction as data subjects is exploited and used for economic gain, in the process reinforcing prevailing inequalities. Finally, its virtuality is reflected in an insatiability matched only by its desirability. There is always more data to mine and no matter how much we 'know' about ourselves, there is always more data to collect and to be judged by. Virtual power is, hence, a simultaneously very real and utterly projected form of control, forecasting and preying upon who we currently are and all the possibilities of who we may one day become.

3
Surveilling Ourselves

The twenty-first century is plagued by what appears to be profound identity crises. Specifically, it is a time when once sacred modern identities are being dramatically eroded, yet the importance of identity has arguably never been so important. When the very foundations for an 'essential' self are being fundamentally challenged, yet individualism continues to reign supreme. This contradictory dynamic, of course, has been much commented upon. It is attributed to a mass sense of loss and the collective need to cling to past truths in a globalising contemporary world. It is an understandable if lamentable psychological response to the deep insecurity wrought by neoliberalism. What is missing from these insights, however, is how accounting technologies and discourses have shaped this present-day search for selfhood. The information revolution and the data economy it has helped spawn have dramatically expanded the possibilities and management of social identity. It has also produced a new form of social power that relies upon 'self-monitoring' to reinforce this more fluid capitalism.

The ability to access and manipulate data, in this regard, has had a massive impact on current identity formation. There has been a veritable explosion of information available that one can use to define oneselves as well as social platforms upon which to do so. Step counters on your smartphone show you and others that you are invested in being an active and fit self. Images on your Facebook account reveal that you are a foodie who loves to socialise with your friends. The profile on your LinkedIn account, by contrast, displays to future employers and professional contacts that you are a professional star. Blogs and tweets also allow you to express your wide range of interests – from politics to fashion – and connect with a diverse set of social networks and users.

At a less conscious level perhaps, data mining reflects interesting truths about your 'real' self that may be starkly different to the way you present yourself. To friends and relatives you may love to pontificate about the latest foreign film. However, your Netflix choices show that you are much more of a sitcom and slapstick comedy lover. A quick search of your recent Amazon purchases and browsing makes it clear that you're mostly interested in new shoes and watches, and not the latest literary sensation or historical tome. The growing prevalence of 'smart' voice devices further allows corporations and governments to capture your daily preferences – potentially revealing to yourself and the world your actual likes rather than what you would like them to be.

There has, moreover, been a distinct existential shift in how we identify ourselves. There is a growing acknowledgement that our identity is not singular but multiple. To this end, we embody not a self but selves. Current theories of intersectionality tap into this more plural understanding of who we are. Here, the imperative is not to discover one's 'true' identity so much as it is to account for the multiplicity of one's social identities as well as their interaction. Digital advances have further reinforced this emerging reality of multiple selves. The ability to connect with others on a wide range of networks under various different guises perpetuates how an individual is less someone and more 'someones'. Put differently, it creates a culture of avatars – a technological version of the old idea of 'one person, many faces'. These exist as diverse sites of identity, virtual places where subjects can try and play out a multitude of roles and ways of being in the world. If nothing else, they provide platforms for experimenting with various self-presentations without any of the risks traditionally associated with identity incoherence.

Indeed, such technology has offered the present generation a fresh comfort with their own and other's subjective pluralism.

However, it also points to a fundamental paradox of contemporary capitalism. As market societies become more unregulated, there is greater demand for identity and selfhood to be properly accounted for. This insight may sound rather strange in an era where relative anonymity and the growing democratisation of media have led to what is commonly perceived as a completely unaccountable culture of trolling and fake news. Yet digging only a little deeper, a more complicated, contradictory and insidious reality begins to emerge. Here there is a need for one to constantly verify and authenticate 'who you are'. Even amid the multiplicity of selves discussed above, while singularity may be on the wane, the demand that these 'persons' account for themselves and even their actions is rapidly on the rise. It is precisely, perhaps, due to the sense of present-day unaccountability that desires for accountability have become so high.

This shift towards a culture of technological and moral accounting has serious implications for the modern evolution of power as it relates to selfhood. The explosion of new data-tracking technology alongside this information revolution has meant that we must now be personally responsible for creating 'smart identities'. These are ones that can tap into diverse networks (whether personal or professional) in order to maximise their value to ourselves and others. While such maximisation is usually more illusionary than factual, it nevertheless stands as an ethical imperative for identity construction. As such, all our diverse selves must be monitored and accounted for, using our enhanced access to their data-driven personal 'histories' to judge whether they are in fact adding to our overall success and well-being as a person. Increasingly this means being subject to a range of external and internal evaluations in order to accurately assess whether these diverse selves are in fact valuable and should therefore be retained.

At stake, therefore, is the production of the fully 'monitored' and 'accountable' self. Selfhood is not so much an essential thing as it is a socially constructed pluralistic entity whose existence depends on its calculated economic and social benefit – its 'added value'. Every action, every remark, every manifestation of self can eventually be made available to such estimations. This represents, moreover, an evolution from self-disciplining based on regulation and governance, to 'self-monitoring' which revolves around creative accounting and market-based accountability. Importantly, this subjective expression of 'accounting' reverses the conventional dynamics of socio-economic accountability. It is now not the economy itself that must be accounted for and judged as to its overall social worth, but the multitude of constantly emerging selves that populate it.

Accounting for a Fluid Existence

There is little doubt that we are living in a simultaneously more connected and fluid world. Globalisation and the technologies that helped make it possible have famously 'shrunk the world'. Digital advances allow people to communicate across previously impenetrable international borders in a matter of seconds. Social media has made virtual interactions and relationships a normal part of our everyday life. Once sacred beliefs and identities are being challenged as perhaps never before.

The late great sociologist Zygmunt Bauman referred to this condition as 'liquid modernity'. Writing at the very beginning of the twenty-first century, he observed that 'These days patterns and configurations are no longer "given" let alone "self-evident"; there are just too many of them clashing with one another and contradicting one another's commandments, so that each one has been stripped of a good deal of compelling, coercively constraining power.'[1] It was a modern existence freed, therefore, in part from its essences. People did not have an identity but rather identities. Their sense of self was multiple and malleable. It shifted with the tide of a rapidly changing world. Hence,

> The liquidizing powers have moved from the 'system' to 'society', from 'politics' to 'life-policies' – or have descended from the 'macro' to the 'micro' level of cohabitation. Ours is, as a result, an individualized, privatized version of modernity, with the burden of pattern-weaving and the responsibility for failure falling primarily on the individual's shoulders.[2]

Leaping ahead only a couple decades, this liquid modernity has evolved into a solidified post-modernity. Identity is now considered by its very nature to be a social construct. It is not a given to be embodied but something which must be continually culturally made up and reinforced. Self-discovery is a matter of constant self-creation. The story of one's life can be told from multiple perspectives and is never straightforward. Emerging theories of intersectionality reflect this fluidity of identity.[3] At the most basic level it asserts the multiplicity of contemporary selfhood. A person is never just one person, they are many combined into one. Their self is socially constructed in accordance to their gender, race, ethnicity, nationality, class, etc. Fundamentally, it asserts the fact that an individual is never exhausted by any single label or version of self.

Yet intersectionality also reveals the deeper tensions of such fluidity. The very multiplicity of identity permits it to be increasingly categorised and therefore accounted for. The age-old question of 'who' we are becomes an accounting exercise in meticulously chronicling our various component selves. It means judging ourselves against key social indicators and ultimately identities. Thus one is a black female urban liberal or a white male rural conservative. The combinations are seemingly limitless, yet they share a seemingly infinite capacity to be documented, indexed and judged accordingly. Consequently, it reveals the complexity of power relations. As Eisner tellingly notes, 'It means understanding that different kinds of oppression are interlinked, and that one can't liberate only one group without the others. It means acknowledging kyriarchy and intersectionality – the fact that along different axes, we're all both oppressed and oppressors, privileged and disprivileged'.[4] Yet it can also, if not properly theorised, focus only on describing these complexities – providing, if you will, a descriptive accounting of these experiences that, while interesting, is far from always being critically illuminating.

Technology has kept apace with this indexible fluidity. It is now increasingly possible to keep track of your various selves as well as how they may intersect and interact. Advancements in big data allows one to investigate their various life possibilities based on their specific identity combinations. Figures on anything from house prices to crime rates to health statistics can be personalised to meet your diverse identity needs. The internet and social media can allow someone to investigate themselves even further, often providing shared experiences from those whose particular identity configurations are similar to their own.

In turn, there is a critical paradox afflicting contemporary identity. The more fluid the self, the more fully it can be tracked and accounted for. The multiplicity of selfhood has become an invitation for it to be continually counted and archived. The self has transformed into an ever-growing plethora of available identities, all of which can be identified and monitored.

Blindly Monitoring Ourselves

The information age has radically expanded the possibilities for identity. Whatever one desires to be, one can find and learn about almost instantaneously. Data on almost all aspects of modern existence is literally available with the click of a button. Reflected at a deeper level are the ways accounting technology has not just reconfigured but also quantitatively enlarged the very scope of modern selfhood. The security once longed for and partially found in singular identities associated with nationalism, religion, class and ethnicity are now being discovered in the construction and accounting of a broad array of personalised selves.

Uncovered, in turn, is a quite revealing tension that goes to the core of present-day identity formation. The more ungrounded selfhood has seemingly become the greater people long and search for it. The theorist Manuel Castells highlights this precise contradiction in his discussion of the network society and identity. Specifically, that the diminishing of traditional identifications, felt to be slipping away as societies become more connected, is met with the inverse popular desire to recover and strengthen them. He observes thus that 'Along with the technological revolution, the transformation of capitalism and the demise of statism we have experienced in the past 25 years, the widespread surge of power expressions of collective identity that challenge globalization and cosmopolitanism on behalf of cultural singularity and people's control over their lives and environment.'[5] The literal and figurative ungrounding of societies from their conventional geographic constraints has led to a resurgent hope that they can be culturally reanchored to a previously secured sense of self.

Identity, in this respect, is intimately wrapped up with the personal and collective need for ontological security. Survival here exceeds simple physical requirements. Instead it involves the situating of oneself in a safe cultural context. Returning again to the insights of Castells, while this desire to push back against this more fluid internationalised world has produced 'proactive movements' such as those associated with feminism and environmentalism, 'they have also produced a whole array of reactive movements that build trenches of resistance on behalf of god, nation, ethnicity, family, locality, that is the categories of millennial existence now threatened under the combined contradictory assault of techno-economic forces and transformative social movements'.[6] These identities serve as an often desperate rearguard defence against the threat of losing oneself – of having no guaranteed place in the world or understandable compass for making clear sense of it.

This echoes Gergen's earlier famous depiction of the contemporary 'saturated self'.[7] Modern technology has, in his view, placed the traditional self 'under siege'. It has created a present-day context where the rise of the internet is leading to disaggregated and disintegrating identities. People are now 'saturated' with so much information and to some extent choice over who they can be, they ironically find themselves paralysed as to actually making this decision and embracing stable self-definition. Quoting him at length, in this respect:

> New technologies make it possible to sustain relationships – either directly or indirectly – with an ever expanding range of other persons. In many respects we may be reaching what may be viewed as a state of social saturation. Changes of this magnitude are rarely self-contained. They reverberate throughout the culture, slowly accumulating until one day we are shocked to realize we have been dislocated and can't recover what has been lost ... Our vocabulary of self-understanding has changed markedly over the past century, and with it the character of social interchange. With the intensifying saturation of the culture, however, all our previous assumptions of the self are jeopardized: traditional patterns of relationship turn strange. A new culture is in the making.[8]

Surveillance and ultimately monitoring is a key part of these contemporary efforts to establish a basic sense of ontological security. It provides a concrete means to regularly reinforce one's identity. The greater ability to collect data about oneself and to monitor its progress is a continual reminder that this is 'who I am'. Counting calories and steps on your smartphone is a daily cue that you long to be a 'healthy' self. Taking pictures of your meals and posting them on social media is a confirmation to yourself and others that you are a 'foodie'. Constantly checking the latest news updates and arguing with people on the internet strengthens your identity as someone who cares about politics and the world. More profoundly, social media can provide an avenue for historically vulnerable populations to 'safely' express their identity – such as the example of young gay men who come out on Facebook and YouTube and in doing so reaffirm prevailing narratives of queerness.[9]

Such accounting practices permit the construction of self to be simultaneously both personalised and marketised. Complex algorithms constantly collect your individual data to cater to your self preferences. A quick Google search of possible holiday destinations can lead to an avalanche of travel ideas and deals. Looking up the score of your favourite sports team on your mobile can lead advertisers to try to sell you their best player's jersey seconds after you discovered if they won or lost. Almost everywhere you look reaffirms your past identity choices and offers you fresh opportunities to recommit to them.

It also significantly expands the possibilities of using these digitally produced identities to achieve a deeper sense of ontological security. Any and all identity is available to be consumed. Even the slightest spark of interest in something can be digitally accounted for and sold back to you as a potentially new identity in which to invest yourself. There is seemingly no limit to ones search for self. The contemporary pursuit of existential and psychic safety is indexible and easily accessed by oneself and advertisers alike. Just as who one is has become multiple, so too are the present-day routes one has to feeling subjectively stable and grounded.

Critically, this reflects an updated version of Foucault's theorisation on the 'technology of the self'. Though often known for his perspective on power and knowledge, in a later lecture he observed: 'Perhaps I've insisted too much in the technology of domination and power. I am more and more interested in the interaction between oneself and others and in the technologies of individual domination, the history of how an individual acts upon himself, in the technology of self.'[10] These technologies centre upon how one is historically socialised to 'take care of yourself'. Foucault notes further that

> There are several reasons why 'Know yourself' has obscured 'Take care of yourself'. First, there has been a profound transformation in the moral principles of Western society. We find it difficult to base rigorous morality and austere principles on the precept that we should give ourselves more care than anything else in the world. We are more inclined to see taking care of ourselves as an immorality, as a means of escape from all possible rules. We inherit the tradition of Christian morality which makes self-renunciation the condition for salvation. To know oneself was paradoxically the way to self-renunciation.[11]

Critically, this also reflects a new perspective for understanding power and control, focusing on 'the ways individuals act on their selves, and how this action on the self can be linked up to actions on the social body as a whole.'[12] This form of power extends to our 'virtual' selves and society. However, present-day practices and values of accounting have once again reversed this dynamic so that 'self-care' is contingent upon 'knowledge' of oneself. The more information one has, the greater one is aware of their preferences and therefore able to pursue them. How we take care of ourselves, thus, is through a continual accounting for our composite selves based on our various tastes and desires. Through such accounting we gain a greater glimpse of 'who we are' in all our personal diversity, and have a better opportunity to tend to these different parts of ourselves. Knowledge is then primarily the personalised data that allows us to explore, expand and care for our possible social identities.

Yet this greater intimacy between identity and accounting does not mean that all is accounted for. What is too often ignored or at least continually put to the side is how confined individuals remain as social subjects. The potential to enlarge one's self has not translated into an equivalent increased capacity to change one's socio-economic situation. Indeed, these opportunities for selfhood have arisen within a neoliberal system marked by rising inequality and downward mobility. Companies, to this end, have become 'entrepreneurs of the self' – crafting employee selfhood to meet increasingly controlling and demanding managerial prerogatives.[13] In this regard, while selfhood is progressively fully accounted for, the capitalist system remains by and large unaccountable. 'Self-expression', as such, has increasingly become an exercise in personalised corporate branding. One large-scale survey of Facebook users, to give one example, found that individuals would 'like' brands that they felt represented their 'inner and social selves', a basis upon which they also formed strong virtual social bonds to others who were similarly 'like' themselves.[14]

Individuals are involved, hence, in a constant process of monitoring one's various identities with rather minimal reflection as to their critical history or present implications. One can delve deep into a film genre without the slightest inkling as to how one is being manipulated by advertising to like certain films over others, or the power to really affect what type of films tend to get made or shown. Identity, as such, has largely evolved into a consumptive activity – a cultural wardrobe to be bought and worn and then disposed of when no longer in fashion or useful. The potential of virtual communities is, accordingly, transformed into digital marketplaces of consumer-to-consumer websites.[15] Technology, therefore, marks out who one bases their existence on as a user. This enhanced visibility can exacerbate existing forms of stigmatisation – such as how those with disabilities can feel even more exposed and trapped in this identity through their employment of readily seen technology that is meant to assist them.[16]

Of course, one should not ignore the political implications of this identity shift. The capacity to collect and share data as well as record real-life events has inspired a range of politicised identities that attempt to deploy this accounting culture in order to make the status quo more accountable. A prime example is the Black Lives Matter protest which has used social media, guerrilla surveillance techniques and data analysis to build a mass movement against racism and police brutality. To this end, the viral virtual handle #BlackLivesMatter was a purposeful attempt to 'move the hashtag from social media to the streets'.[17] Nevertheless, these types of collective physical struggles are becoming in many ways the exception not the rule, as digital media technology has

> given rise to an era of personalized politics in which individually expressive personal action frames displace collective action frames in many protest causes. This trend can be spotted in the rise of large-scale, rapidly forming political participation aimed at a variety of targets, ranging from parties and candidates, to corporations, brands, and transnational organizations. The group-based 'identity politics' of the 'new social movements' that arose after the 1960s still exist, but the recent period has seen more diverse mobilizations in which individuals are mobilized around personal lifestyle values to engage with multiple causes such as economic justice (fair trade, inequality, and development policies), environmental protection, and worker and human rights.[18]

The danger is how easily these data-driven forms of lifestyle politics can be manipulated to serve the powerful. It is not just the age-old adage that statistics can be used to prove anything. Instead it reflects the greater ability to use data to create highly politicised identities with little regard to either accuracy or justice. Information is propagated that paints an inviting picture of alternative realities with victimised selves to invest in that are as reactionary as they are insidious. Returning to the example of Black Lives Matter, this movement helped to fuel racist discourses and identities that reinforced white privilege, spurring the renewal of explicit white nationalism while strengthening an authoritarian police culture.

What is overwhelmingly present is a culture of individuals and communities blindly monitoring themselves. More precisely, while some people and communities have used this new technology to help become 'woke' (a contemporary phrase referring to an individual becoming more aware and sensitive to prevailing social and economic injustice), by and large there is a culture that increasingly accounts for itself without deeper reflection or the capacity to change the larger socio-economic system producing these selves. Two points are particularly relevant for the analysis here. The first is how these accounted-for identities do not lead naturally to a culture of greater accountability. By contrast, it often deploys accounting technologies and an ethics of accountability to allow those in power and the system itself to be unaccountable. The second is that ontological security and truth are less often found in any conventional notion of essence – conversely it is linked to the prevailing call to use accounting to ensure one's own accountability. At stake are new ways of controlling the subject and establishing social domination – themes that will be explored in greater detail in the following sections.

Smart IDs

Identity formation is becoming increasingly sophisticated and accounted for. The possibilities for selfhood have exploded over only the past few decades. We now have the information to be seemingly anyone or anything that we desire. One can look up how to cook Chinese noodles in the morning, the latest heavy metal band in the afternoon and the health of the stockmarket in the evening. An initial point of caution here may be that tastes do not make identity – they may form its ingredients but are not exhaustive or completely reflective of who one is. However, in the contemporary age as the idea of an inherent essence has retreated, 'who one is' is more and more a tallying up of one's digitally collected interests and preferences. It is a private and public collection of the things a person has explored and done virtually – personal data that can be regularly archived, reviewed and mined for both identity and profit.

This points to a broader evolution in how selfhood is experienced and expressed in the shift from modernism to post-modernism. As highlighted throughout this chapter, the previously secure foundations of the modern world have liquefied considerably. Whereas once almost unquestioned points of identity such as nation, class, race and religion largely determined one's sense of self, these categories have now become fluid and far from overdetermining. There is an even more dramatic change occurring as well: the very story of oneself is being radically altered and retold. The straight-ahead chronological narrative detailing a person's life from birth to death is being augmented and to a certain extent supplanted by something decidedly more fractured, and to most traditional points of view incoherent. This resonates with a post-modern ethic where the straight-ahead narrative is displaced by something considerably less coherent and linear.

That the self would expand and fragment is perhaps not that surprising in light of the general death of ideology.[19] Past steadfast and unending beliefs in the truth of communism, fascism, even liberal democracy have waned or disappeared almost altogether. In their place is a much more flexible sense of self – one open to opportunities, able to move easily between belief systems when desired and adaptable to whatever is trending. However, what was perhaps far less predictable was how monitored and accounted for this post-modern self would become in practice. If modernity has in fact been deconstructed, it has also been reconstructed as a post-modern reality marked by enhanced surveillance, data collection and a permeating ethos of constant personal accounting. While not every story can be told, every moment can be potentially captured and codified as data for present consumption and future use.

Obviously this is not the whole picture. There is a considerable modern reinvestment in what now has become the conventional self. Barber's famous early discussion of 'Jihad vs. McWorld' pitting fundamentalism verses corporate globalisation exemplifies this complexity.[20] This also applies to a range of conventional modern ids that are not necessarily extremist in nature (or at least are not conventionally assumed inherently to be so). There has been a renewed embrace of modern identities such as patriotism and religious devotion. Nevertheless, this modern resurgence has a distinctly post-modern flavour, one centred on the values and practices of accounting. Nationalism and traditional family values are now less a concrete way of life and more an ideal 'lifestyle' and set of beliefs that one defends and posts about on social media, as well as a collection of purchasing preferences. Thus people post on Facebook that they are disgusted that an athlete refused to stand for the national anthem, shares that they went to church today for their 'friends' to see and then tries to find where the latest Christian film is showing.

A key feature of present-day selfhood is the use of accounting to cultivate smart identities. 'Smart technology', in this respect, allows people to become better and more informed versions of themselves. If you want to unleash your inner gardener, one can look up the best techniques, ask other green-fingered folk around the world for advice, blog about one's challenges and triumphs and even download an app to record your progress. Identity is now an intimately accessible and perfectible experience. Contemporary accounting technology and practices provide people with the opportunities to create selves that are 'smarter' than seemingly ever before. They serve, in the famous phrase coined by MIT scholar Sherry Tuckle in her landmark book *Life on the Screen*, as 'identity playgrounds' where people can use the virtual world to try on different identities, many of which stand in stark contrast to their offline self.[21]

Indeed, this new online reality doesn't only provide a space for self-experimentation but also for profound self-improvement. This prevailing 'smart culture' represents a fresh way for individuals to engage in self-improvement. All of these accountable selves are a snapshot of where one ultimately desires to be – whether that be the ultimate professional, the most informed political commentator, the pre-eminent concert goer or the most successful dater. It is precisely here that accounting and accountability presently intersect. A person's smart identities are constant external reminders of their imperfections and their need to be better. Significantly, this ethos of continual improvement must be a two-way street, so that technology accommodates the needs of different users in 'helping them help themselves'. One study thus showed that fitness apps could do substantively more to assist older users through such measures as using bigger fonts and introducing smaller target sizes.[22] Nevertheless, this reveals the broader association of being 'smart' with ideals of bettering and perfecting oneself.

This insight echoes and builds upon the ideas of identity work.[23] This concept describes 'people being engaged in forming, repairing, maintaining, strengthening or revising the constructions that are productive of a sense of coherence and distinctiveness'.[24] In the post-modern world, this work has been reformatted. It reflects the more fragmented and fluid character of contemporary identity, a reality captured in notions of *identity bricolage* where people 'cobble together' a sense of self based on their diverse identities.[25] However, now this 'work' is undertaken through the use of smart technology. It is a regularly updating referendum on your progress to becoming a perfect self in whatever way you seek to be. It is a small audible ping from your pocket or purse that rings loudly in your mind asking if you met your daily step goals. It is the buzz in your hand that briefly jolts your consciousness reminding you that you are late for a date with a friend.

Information technology has, in this respect, begun to irreversibly alter the very configuration of identity. It is no longer founded solely, or even primarily, on conforming to the cultural norms and expectations of people in 'real' life. Rather it is premised on processes of constant virtual verification and validation. Positively, people draw on 'information technology artefacts' such as a digital history of their past interactions to reinforce their sense of identity and actually contribute to the knowledge of these online communities.[26]

Digging beneath the surface of identity, this 'smart' accounting is fundamentally reloading contemporary selfhood. It is not just that it verifies who one is, it also continually validates that they are someone in the first place. It reflects a new era of the self whose existence is formed and made possible through external data collection and digital self-presentation. Hearn, for instance, has recently revealed the disciplinary effect of virtual 'identity badges' driven by big data, such as the Twitter verification checkmark. While seemingly innocuous, they in fact exist as

> both an affective lure that incentivizes specific styles of self-presentation and a disciplinary means through which capitalist logics work to condition and subsume the significance of the millions of forms of self-presentation generated daily. Beneath the promise of democratized access to social status and fame, the business practices of the social platforms in and through which we self-present draw us into privatized strategies of social sorting, identity management, and control (published online).[27]

Our reasonable concerns over identity theft reveal a more fundamental existential insecurity. 'Who am I but my digital footprint?' is the underlying theme of the age. The fear is that if one's data are erased so to will they be. Deleting an individual's digital information is akin to deleting them entirely.

The attraction of this technology is its ability to easily and continually allow people to account for themselves. By simply turning on one's phone, an individual reconfirms that they exist. It is not surprising, in this light, that people so often personalise their phones – it is more than an expression of their identity, it is a declaration that they are in fact a real self. To paraphrase Descartes for a new age, I text therefore I am. Revealed is a present-day self that is as fluid as they are accounted for. Their embrace of multiplicity is transformed into different codified and categorised selves. New technologies have created fresh wisdom for making them, furthermore, the very best selves they can be. The world is now full of people loading up smart technologies so that they can embody smart identities in real time. In accounting so intelligently for themselves they have also become increasingly accountable to contemporary capitalism.

Rating Yourselves

The creation of smart identities is meant to be at the cutting edge of personal and collective empowerment. It permits people to download a new life and upload fresh possibilities for self-creation. It allows individuals and communities to see through the modern matrix of post-modern existence. It transforms its fluidity into a concrete dataset personally formatted and archived to match your diverse preferences. In this regard, these smart identities not only provide the opportunity to discover your various selves but also constantly improve upon them. However, in practice these identities are often more regulative and focused on creating the perfect capitalist self than they are emancipatory and full of genuine possibility.

Indeed, this smart culture reveals the evolving ways post-modernism and neoliberalism are intersecting and reconfiguring present-day selfhood. Neoliberalism is associated with the growing marketisation of society. All spheres of life can now be bought, sold and exploited for maximum profit. Just as significantly, this market logic is increasingly shaping current rationality and desires. To this effect, popular self-help books like Steve Covey's *7 Habits of Highly Effective People* in fact represent

> epiphanogenic (or epiphany inducing) technology emerging from an 'effectiveness' discipline supported by three socio-cultural trends: the postmodern, saturated self; the coming of neo-liberal society and the financialization of the self; and the subjective turn. Covey's discipline of effectiveness aims to produce a self that is simultaneously de-saturated, financialized and expressivist, but supportive of conservative, universalist and late capitalist modes of being.[28]

Such smart accounting is absolutely central to this complete capitalist take over of the self and identity. The ability to collect and analyse personal data turns identity formation into a constant calculation of one's overall efficiency and value. Smart values are compatible with and in fact mutually reinforce these market prerogatives. Consequently, the use of smart technology such as mobile phones has transcended mere person-to-person communication and now serves as a broader device for wholesale 'identity management'.[29]

Going even further, techniques like people analytics can tell companies with increasingly exact accuracy just how uniquely valuable each of their employees is. To this end major companies like Google are using 'data-based people management' on a 'quest to build a better boss'.[30] Using sophisticated data collection and analysis methods, these techniques pinpoint where individuals, groups and organisations can become more efficient and productive. It is trumpeted as a hi-tech, cutting-edge way in which 'Advanced analytics provides a unique opportunity for human-capital and human-resources professionals to position themselves as fact-based strategic partners of the executive board, using state-of-the-art techniques to recruit and retain the great managers and great innovators who so often drive superior value in companies.'[31] Their purpose is to in a sense uncover where there are gaps in 'intelligence' so that people, places and things can effectively maximise their goals and therefore value. Critically,

> Big Data continues to be touted as the next wave of technology and analytics innovation. From our perspective, the next wave of innovation is not just about Big Data, but more about how companies leverage Big Data analytics to take action and optimize their business. Having data is not enough; it needs to be leveraged effectively to drive and optimize business action that is coordinated at all levels of the organization. As it relates to People Analytics, Big Data is critical to providing real-time insights to businesses regarding how to maximize the value of the talent for the organization as well as maximize the organization's value for the talent it intends to retain and develop.[32]

Interestingly, people analytics is commonly linked to the improvement of well-being. Through better understanding of how one works, lives and plays it is possible to judge if they are maximising their time successfully. Of course, these claims can ring hollow for a growing number of people in an age of 'time-greedy' organisations.[33] Yet these rather empowering aspirations, even if they are only rhetoric, reveal the depth to which this market-based monitoring logic has colonised current thinking and desires. The discovery of 'smarter' ways of doing things – meaning how to do more with less – is the key to achieving all your hopes and dreams.

To this effect, it is now possible to rate all our actions and principles as to how smart they are. Are you deploying the best, most efficient and least resource-intensive strategy for attaining your goals? What is emerging is a culture of constant external and internal regulation – in which one's data are the basis upon which their worth is ultimately judged. It also makes one not only an active consumer of information but also an active producer of it. Hardey, for instance, writes of how those who have suffered an illness will use personal webpages to publicly tell 'the story' of their ordeal and in doing so transform themselves from 'consumers of health information and care to producers of information and care'.[34]

This brings new meaning to the phrase 'self-management'. This concept has been a centrepiece of neoliberalism – preaching the capacity of individuals to monitor and discipline their own conduct in line with market expectations and demands. To this extent, it is imperative for technology to become smarter so that individuals can become smarter. The goal, thus, of the much-heralded 'internet of things' is to reach a point in the near future 'where intelligent devices operate in concert to enrich the overall user experience by sharing resources and capabilities'.[35] Consequently, now simple self-management is not enough. What is required is 'selves-management'. Notably the ability to personally ensure that one's various identities, both individually and together, are 'smart', productive and profitable. Arising are new apps that promise to help people achieve 'work–life balance' through 'smart' features such as 'task collaboration', a 'family to do list' and even a 'sleep cycle alarm clock' that helps to regulate your rest. There are also apps that aim to optimise individuals' productivity with such revealing names as 'coach.me' and the 'focus booster' that will assist you in maximising your personal and professional life in the long term and on a daily basis, respectively.[36]

It also therefore reconstitutes the very definitions of the work and life supposedly being balanced. In the first place, there is increasingly no such thing as a non-working life. All of one's experiences should 'work' to improve their well-being and life prospects. It is imperative, in this sense, to live 'smartly' no matter what you do. Just as significantly, people progressively have more than just one life – rather they lead lives in the plural. As such, conventional desires for balance increasingly revolve around balancing these lives 'intelligently' and efficiently.

This attempt to hold one's selves accountable reflects theories of identity regulation. Akin to the previously discussed concept of identity work, such regulation denotes how prevailing identities can shape and come to dominate subjectivity and selfhood.[37] It highlights how 'organizational control is accomplished through the self-positioning of employees within managerially inspired discourses about work and organization with which they may become more or less identified and committed'.[38] The neoliberal injunction to be 'smart' is a powerful current discourse for governing the individual and collective self to reflect capitalist and corporate principles and desires.

These insights represent, in turn, an almost complete reformatting of established sociological accounts of the self. Arguably one of the best known and still most relevant is Goffman's idea of a front and backstage self.[39] In the age of smart accounting and accountability, it is perhaps more accurate to speak of front and back operating platforms. Operating platforms refer, in this regard, to the various social media outlets that one is present on and equally presents themselves on. These serve as new front stages for people to act out their preferred self based on perceived social expectations and personal interaction. To this effect, each site requires individuals to 'smartly' tailor themselves to its specific cultural specifications. An early study of Facebook users, for instance, found that 'users predominantly claim their identities implicitly rather than explicitly; they "show rather than tell" and stress group and consumer identities over personally narrated ones'.[40] Another interesting and more recent example is the rise of the 'quantified self' in the health sector, where individuals embrace self-tracking technology to enhance their physical well-being. While on the surface this may sound like an advance for personal and public health, in practice it commonly prioritises 'the visible and metric' as opposed to deeper and less immediately seen symptoms.[41] The backstage is, hence, the hidden programmer, keeping track of and managing these self-presentations. In this spirit, Gardner and Davis depict the rise of the so-called 'packaged self', highlighting the more externally focused identity of the current app-based younger generation and their desire to effectively sell their visible digital 'self' to other users.[42]

This emerging practice of personal self-regulation is easily uploaded and transferred to the workplace. The formerly strict demand to embody corporate values is fading away and being replaced by an ethos of 'just be yourself'.[43] Yet this allowable freedom is intertwined with an equally strong culture of contemporary accounting and accountability. Specifically, you can be whoever you want (within corporate-approved limits) just so long as you can show that whoever you choose to be is productive, efficient and ultimately profitable. This is especially evident in the rise of the precarious and freelance economy that has accompanied the growth of neoliberalism. In a time where the traditional employment biography is seemingly dying, the ability to deftly regulate one's multiple identities comes particularly handy as it permits one to meet the malleable needs of their ever-changing employers.[44] One can quickly morph into the perfect employee for a specific project and client. With zero-hour contracts, this adaptability of one's selves is at a similar premium. Temporary jobs means creating temporary and flexible working selves. The ability to bid for positions and appeal to employers is about 'smartly' accounting for the employee they desire and fitting your own self to these criteria.

Consequently, at the heart of neoliberalism is a profound emphasis on rating one's selves. All of an individual's identities are indexible and available to careful and considered scrutiny. A pressing question is how valuable is this identity for me? Does it serve me well or should it be fired and deleted? These determinations are made by the judgements of bosses past and present to fulfil one's existing and future capitalist needs. In turn, the evolution from disciplining to monitoring as the primary means for governing and controlling the self is revealed.

Producing the Self-monitoring Subject

Current monitoring technologies reflect the profound conundrum of the contemporary self. There have perhaps never been so many identities for individuals to choose from. Smart technology and social media have made this embrace of multiple identities not only possible but normal. Yet it is precisely this technology that has also made these selves so indexible and ultimately controllable. People are asked to at once account for their various selves and ensure they are accountable to the market demands of neoliberalism, which reveals the proliferation of an evolved means for producing and managing the present-day capitalist subject.

What emerges, in turn, is a type of self that is seemingly infinite in its potential manifestations, yet decidedly restricted in its actual possibilities due to its heightened ability to be monitored and shaped by existing power relations and discourses. Foucault's notion of self-disciple goes a long way in helping to illuminate this apparent contradiction. He declares that 'discipline may be identified neither with an institution nor with an apparatus; it is a type of power, a modality for its exercise, comprising a whole set of instruments, techniques, procedures, levels of application, targets; it is a "physics" or an "anatomy" of power, a technology'.[45] Self-discipline represents, therefore, the diverse norms, institutions and other everyday social forces that shape the knowledges and practices of the self.

An immediate objection to simply equating monitoring with discipline, is that it fails to capture just how empowering and creative accounting for ourselves can be for contemporary subjects. It is as much an opportunity for personal expression as it is a perceived threat to their autonomy and freedom.

Indeed, Foucault himself points to this tension between how power simultaneously expands and limits the social potential of subjects. He distinguishes between 'the economic' and 'the political' for this purpose:

> discipline increases the force of the body (in economic terms of utility) and diminishes these same forces (in political terms of obedience). In short, it dissociates power from the body; on the one hand, it turns it into an 'aptitude', a 'capacity', which it seeks to increase; on the other hand, it reverses the course of the energy, the power that might result from it, and turns it into a relation of strict subjection. If economic exploitation separates the force and the product of labour, let us say that disciplinary coercion establishes in the body the constricting link between an increased aptitude and an increased domination.[46]

Similarly, monitoring grants people new techniques for developing their selves while also delimiting it to the narrow horizons of a free market discourse. It is, thus, at one and the same time an expansive economics and restrictive politics of the present-day self.

This dual aspect of selves is witnessed in the modern 'empowerment' of employees. Indeed, even in the face of growing economic precarity and inequality, we have supposedly entered the 'empowerment era'. Here organisations are expected to help their members fulfil their personal and spiritual needs as well as their economic ones. These new 'human-centred' organisations aim to increase the 'physical and mental health of employees', including their 'advanced spiritual growth and enhanced sense of self-worth'.[47] Yet such empowerment often has quite insidious consequences, leading to greater work intensification and in some cases increased anxiety. According to Willmott, 'Corporate Culturists commend and legitimise the development of a technology of cultural control that is intended to yoke the power of self determination to the realization of corporate values from which employees are encouraged to derive a sense of autonomy and identity.'[48] Particularly relevant to this analysis is how these empowerment values and practices serve simultaneously to 'economically' expand and 'politically' limit the contemporary subject.

Represented is a Janus-faced existence between possibility and restriction in present-day selfhood, especially as it relates to themes of accounting and accountability. MacLullich's fascinating study of the introduction of more technologically sophisticated 'auditing regimes' points to this paradoxical relation. He observes how these 'new strategic audit' discourses only provided 'the appearance in change' as the 'sophistication of programmes and the appearances of professionalism delimits the amount of time available for the exercise of judgement and interpretation in the audit process'.[49] The potential for self-expression has almost undeniably been enlarged in terms of preferences and constricted in terms of political and economic agency. You can, so to speak, be anything you desire just so long as it is profitable – or at the very least not unprofitable.

This reflects how discipline thus forms only one part of virtual power. It undoubtedly seeks to contain and 'fix' subjects in line with what would be expected of disciplinary regimes in the Foucaultian sense. However, it also expands the scope of market discourses for configuring selfhood. Such virtual monitoring has made it so that every activity, identity and expression of self conform to a capitalist logic of efficiency and maximising value for resources. It provides the material and virtual resources for, according to Gill, 'managing the self in an age of radical uncertainty'. Specifically, in her view,

> new media work calls forth or incites into being a new ideal worker-subject whose entire existence is built around work. She must be flexible, adaptable, sociable, self directing, able to work for days and nights at a time without encumbrances

> or needs, must commodify herself and others and recognise that – as one of my interviewees put it – every interaction is an opportunity for work. In short, for this modernised worker-subject, 'life is a pitch'.[50]

All selves are, in this spirit, indexible and judged according to their financial worth. Through such accounting one can constantly assess how they responded to these constantly appearing 'market opportunities', and whether they effectively took advantage of them in the most optimal way possible.

In such a situation, a novel form of social power driving and shaping selfhood is present. Self-discipline has been updated to self-accounting. More precisely, it is the creative and expansive ability to create 'smart' market selves. All possibilities regarding who one is and would like to be must be fully accounted for and accountable to the larger demands of efficiency and profits. Underpinning this power is an entire cultural system designed for this purpose. From smart technology to social media to big data, everything is oriented to encouraging subjects to craft valuable market selves. The greatest production of contemporary capitalism is ultimately ourselves.

Investing in Ourselves

The present period has seen the rise of the self-accounting subject. Individuals now have the technology to create ever newer selves. Yet these possibilities are increasingly monitored, categorised and technologically accounted for. Even more critically, they must be constantly accountable to capitalist demands to be efficient and profitable. Just as significantly, this constant process of self-monitoring is fundamentally reshaping contemporary subjectivity – recasting our desires to reflect this simultaneously expansive and restrictive market ethos. Specifically, it revolves around the cultural fantasy of the 'fully accountable self', offering a novel affective discourse incentivising individuals to at once psychically and economically 'invest in ourselves'.

What emerges is an appealing contemporary vision of the self linked rather ironically to virtual power. It critically puts in stark relief its profound subjective impact. The psychoanalytic theories of Jacques Lacan capture this affective dimension of virtual power. His conception of fantasy is especially pertinent in this respect.[51] Rather than its popular connotations as a type of illusion, fantasy here depicts the cultural ideals that we strive to personify, and in doing so achieve an always elusive sense of psychic harmony. Quoting Žižek, it represents 'the bonds linking together its members always implies a shared relationship to the Thing, toward enjoyment incarnated ... If we are asked how we can recognise the presence of this Thing, the only consistent answer is that the Thing is present in that elusive entity called our "way of life"'.[52] Importantly, it is not the attainment of this fantasised self that is central, but instead its eternal pursuit. For this reason, Stavrakakis refers to it as a 'failed identification', as 'for even the idea of identity to become possible its ultimate impossibility has to be instituted. It is this constitutive impossibility that, by making full identity impossible, makes identification possible, if not necessary'.[53]

In the current context, the romanticised big Other upon which selfhood revolves is the subject who has fully accounted for themselves and in doing so maximised the value of their selves. At the most basic level it helps alleviate the anxiety created by this seeming technological takeover of all facets of modern existence. In this fantasy, it is us not our phones, computers or big data that is in control. Looking at individual perceptions of identification and authentication technology, in this respect, Zoonen and Turner observe that

> People experience little problems with their current means of I&A [identity and authentication] and do not like the kind of futuristic means of I&A that are presented in popular

> culture, arts and design, and some R&D departments of big corporations. If people see room for improvements of their future means of I&A, they tend to desire higher ease and transparency of the cards they use. People hope and expect I&A in the future to become even more personalized; they hope to get more control over their online identities but there is widespread doubt this will become possible; they fear and expect commercialization of I&A services, and expect that surveillance will expand.[54]

The overriding desire is that we are able to account for ourselves instead of merely being technologically accounted for.

Fundamentally, it is a crucial way for individuals to seek to overcome their present-day experiences of alienation. Traditionally, this implied the existence of a 'genuine' or essential self that was being suppressed by social forces. However, the self-monitoring subject completely reconfigures this dynamic. Here it is not about maintaining an essential sense of 'who I am' so much as it is a struggle for who shapes and gets to manage this selfhood. There is of course no singular 'me' – to paraphrase Whitman for the present era we are all now 'multitudes'. Rather, it is an internal and external struggle to feel that we are guiding these smart identities rather than merely being at their mercy. Costas and Fleming point to this evolution in the experience of alienation in which they begin to feel as if they are 'strangers to themselves'.[55] What is felt to be lost and must be protected for present-day subjects is not any inherent self in the conventional sense but rather a core part of who they are that has eluded complete socialisation.[56]

The embrace of 'smart identities' and that fantasy of the fully accounted self that underpins them is a further reflection of this attempt to escape alienation in the (post-)modern age. It is the promise that by mastering these technologically driven techniques one can achieve mastery over themselves. Michael Zimmerman philosophically explores this very tension in a piece whose title asks if we have reached 'The End of Authentic Selfhood in the Post-Modern Age'. He begins by affirming that the self has in fact been 'decentred', a change that has been as much liberating as dehumanising. 'Many people find themselves confronted with captivating, expansive and seductive options that allow people to readily exchange one identity for another, such as Internet chat rooms', he observes, noting further: 'That people relish the freedom to explore new technologically generated options and alternative social identities is evidenced by the vast sums of money being spent on them. Yet despite all the excitement some people report feeling disintegrated, superficial, even dehumanized.'[57]

This points to a potential 'technological nihilism' predicted by Heidegger, where individuals are simply 'flexible raw material for a technological system'. However, Zimmerman continues to hold out hope for the potential of authenticity, noting that the anxiety produced by these technologies creates the pretext and desire for subjects to continually 'choose' one possibility – though across a multitude of selves. While undeniably interesting, this analysis also gestures towards the fantasy of 'self-mastery' associated with accounting in this technological age. Notably, it is a mastery accomplished through the use of data and digital communication to personally 'track' and archive oneself, and in doing so having the information to 'choose' who one prefers to be in an uncertain world.

Such processes of self-accounting through datafication become an appealing pretext to pursue multiple versions of oneself at once, without surrendering to a life fully determined by the hidden algorithms virtually surrounding us seemingly at all times. Almost perversely, it is exactly this contemporary form of hi-tech surveillance that contributes to their sense of empowerment. It offers individuals the opportunity to not merely navigate but 'take control' of their lives. Indeed, the more fully accounted for they are, the more empowered they often feel they can be.

These insights reconfigure the increasingly prevailing ethos of 'self-management'. At play is much more than a contemporary managerial imperative to simply regulate one's conduct. Rather, it is an expansive call to explore diverse identities while governing them in such a way to always add value to one's overall existence. Self-management is transformed into a more creative process of 'selves-management'. This reflects an emerging desire to link every and all identity to increasing one's employability, a call for individuals to 'pre-occupy the self with the self'.[58]

This reveals, in turn, how such an attractive, affective discourse of being fully monitored intersects with renewed demand for capitalist accountability. One must at all times shape their identities to meet the diverse and ever-evolving needs of the marketplace. This is readily witnessed in the fantasy of employability pervading contemporary economic culture.[59] Here, one can never be employable enough – there are always selves to develop and existing selves to improve. Freedom is associated almost inexorably with employing accounting techniques to become constantly accountable to employers.

Importantly, while this deep accounting/accountability dynamic creates a palatable mass anxiety, it also produces a fresh – though eternally disappointing – form of subjective empowerment. Specifically, it infuses people with an entrepreneurial spirit that prophesies their ability to deploy their skills and diverse selves to control their own destinies and make a lasting impact on their community and world. Significantly, this combines a profit motive with a fleeting psychic and ontological security. It is the pursuit of this ideal, one whose entire possibility is premised on being more accounted for and accountable, that drives and stabilises selfhood. Employability, consequently, serves to 'indicate how people should behave and what their responsibilities are'.[60] In doing so, it supposedly gives them the knowledge and tools to 'take control' of their professional destiny.[61]

What accountability thus effectively offers subjects is the opportunity to constantly invest in themselves – both psychically and economically. It is an empowering but elusive fantasy of being fully accounted for and in thus 'smartly' governing their own lives. Technology is rebooted from a force of subjection to one of subjectification – in which the continuous and evolving culture of being permanently monitored, analysed, categorised and datafied is perceived as an opportunity to shape their own identity and personally maximise their market value. This investment, even when profitable in the traditional sense, always brings the diminishing returns of a capitalism that unaccountably rules our lives.

Monitoring Ourselves

This chapter has highlighted how the contemporary self is increasingly digitally monitored and made accountable to the free market. The hi-tech smart technologies that have come to largely define this era reinforce the deeper social technologies of self-monitoring that ultimately sustain it. The expansion of the virtual and physical possibilities of present-day identity are confined to a narrow version of a 'valuable' neoliberal self. Across the seemingly ever-growing potentialities of expressing 'who I am' is a universal demand to be efficient, productive and profitable. Not surprisingly, perhaps, a profound by-product of this increased personal accountability is an equally dramatic increase in the unaccountability of capitalists and capitalism.

At perhaps the most simple, though no less important level, this culture of constant monitoring has rather ironically not extended to capitalist elites. Certainly, high-profile politicians are progressively scrutinised as digital technology often brings their past statements, votes and behaviours back to haunt their present ambitions. However, this type of vulture politics pales in comparison to the overall free pass given to the wrongdoings of executives and political leaders. The 2008 financial crisis revealed the underlying corruption underpinning contemporary neoliberal economy and society. In the subsequent years of 'recovery', the idea that the system is rigged to benefit the '1 per cent' while leaving the '99 per cent' behind has justifiably grown. This reflects a distinctly classist system of accounting and accountability – where those at the top have relative immunity and those in the middle and bottom are progressively monitored and held to account.

This unevenness in accountability raises even more fundamental questions of what is being monitored and for what reasons. Tellingly, while there are more data available than ever, our political and social imagination seems to have steadily declined. Bauman hints at the reason for this paradox in his description of liquid modernity. He observes:

> The overall order of things is not open to options; it is far from clear what such options could be, and even less clear how an ostensibly viable option could be made real in an unlikely case of social life being able to conceive it and gestate. Between the overall order and every one of the agencies, vehicles and stratagems of purposeful action there is a cleavage – a perpetually widening gap with no bridge in sight.[62]

Creativity is almost exclusively directed towards expanding one's life choices and identities. The ability to change the system or the agency to conceive of a totally different social order is considered fantastical, while the ability to 'smartly' create and produce new capitalist selves is encouraged and celebrated.

Indeed, even the information that is collected assumes the permanency of the market as if it were akin to an article of faith. Big data and analytics focus primarily on how to maximise consumption and efficiency, respectively. There is seemingly little interest in the ways non-market organisations and practices can provide a viable and preferable social alternative. Even the rise of the 'sharing economy' focuses on the ability of people to find new ways to profit from a 'post-employment' economy. Here, all manifestations of the self are meant to exist with a relatively unalterable and simply taken-as-given capitalist reality. Ignored are emerging ideas, by contrast, of 'sharing cities' that eschew this market fundamentalism, proposing instead

> understanding cities as the political, economic and cultural drivers of global society, thus linking the sharing of urban spaces with the sharing of global resources. It also means understanding cities in themselves as shared entities with shared public services ... shared infrastructure ... and shared spaces. But we go still further in seeing not only a 'right to the city' and to the 'urban commons' ... but also a right to remake them.[63]

This repression of a more expansive vision of 'smart progress' extends to what is deemed valuable. Here what is worthwhile is ostensibly associated with personal fulfilment. Yet in practice this translates to pursuing activities that 'add value' to your life. More precisely, the capacity to use data and technology to optimise the benefits of one's preferences and chosen activities. Hence, according to Spicer and Cederstrom,

> Today wellness is not just something we choose. It is a moral obligation. We must consider it at every turn of our lives. While we often see it spelled out in advertisements and life-style magazines, this command is also transmitted more insidiously, so that we don't know whether it is imparted from the outside or spontaneously arises within ourselves. This is what we call the wellness command. In addition to identifying the emergence of this wellness command, we want to show how this injunction now works against us.[64]

Deeper existential questions of the worth of the market or capitalism are, by contrast, left largely unasked. Indeed, all preferences are accounted for except for the choice over the very social and economic system in which one is forced to lead their life and give birth to their self.

The present, therefore, is a form of constant self-monitoring that masks the deeper unaccountability of contemporary capitalism. Selfhood is turned into a continual journey of personal data mining, assessment and judgement. Here an individual is often their own judge, jury and executioner. Like scrutinising lawyers people meticulously study the digital evidence available to determine the degree of their guilt and whether one of their selves can be 'smartly' rehabilitated or must be terminated. What is so often not judged or accounted for is the capitalist system responsible for so much of their anxiety, and the daily and wider oppression surrounding them. In being forced to virtually account for ourselves the broader global reality of capitalism escapes our attention and governance. In the present period it is not 'care for yourself' or even 'know yourself' that truly matters, but rather 'monitor yourselves'. And it is in such monitoring that our ignorance of our broader world and incapacity to fundamentally shape it festers and grows.

4

Smart Realities

If there is one supposedly universal feature of the current era, it is that everyone is now living in a capitalist world. The once rather defined space of the marketplace has spread to all corners of the globe. Across geographic, cultural, ethnic and class divisions there is increasingly a shared condition of capitalism. Yet just below the surface of this apparent total victory of the free market is a substantially more complicated and less solid reality. New technologies have blurred the line between the virtual and the real, expanding and to an extent complicating the very notion of traditional space. Indeed, even people who live and work in close physical proximity to one another often inhabit profoundly different 'worlds' – coexisting with one another while being part of quite diverse digital networks, engaging in alternative lifestyles and exposed to contrasting sets of information.

This fluidity also seems to hold true for modern times. Flexible employment and smart technology are increasingly making how one passes through life as much a personal lifestyle choice as it is a one-size-fits-all form of social regulation. Indeed, we are progressively our own timekeepers and schedule makers. These contradictions raise serious questions regarding the assertion that capitalism is taking over the world, now largely accepted as a point of faith. Notably, what precisely is capitalist space and time in this all-pervasive global market reality?

On the face of it these should be relatively easy questions to answer, yet they yield surprisingly complicated and at first glance unclear results. In point of fact, a world that is meant to be completely commodified and easily calculated is quite hard to fully quantify in any straightforward or obvious way. The advances of social media and big data certainly provide a literal and figurative wealth of information. However, it is always open to interpretation and eternally incomplete. There are always more data to gather, more findings to analyse and debate. Similarly, no space is ever fully complete nor any time ever completely exhausted. Any place can be used better and the time within it spent more wisely. Hence, the reality of capitalism is as much an ideal as it is a concrete reflection of the 'real' world.

Yet it is precisely this virtual contradiction that helps to power the contemporary free market. It is the productive tension between the growing technologies of quantification and the fact that our lived realities can never be fully quantified that drives twenty-first-century capitalism forward. Every space and every action avails itself to data collection and analysis. It is an information-driven culture that must constantly update itself – understanding, reinterpreting and then making valuable ever new sets of experience. Our present actions do more than shape our future outcomes; they form the very basis for predicting what we will do in the future and how we can do it more effectively. Monitoring here becomes a never-questionable urge to quantify our communities, our world and ourselves. It is only in doing so that we can truly make our environment and time worthwhile.

What is crucial, in this respect, is that the productive capacity of capitalism has shifted from manufacturing goods to manufacturing realities. The accounting revolution, once motored by quantifying technologies both scientific and social in character, do not simply extract data from an 'objective' world. Instead they help to guide, mould and indeed produce them. Their purpose is transformational, turning existing places, people and things into more efficient, productive and profitable parts of a constantly updating and expanding collection of market environments.

Time and space are simply raw materials for the creation of quantifiable and therefore accountable marketised worlds.

Marx famously referred to capitalism, as discussed in previous chapters, as fundamentally 'insatiable' and 'rapacious' – unquenchable in its thirst for fresh markets and labour to exploit. Traditional colonialism is framed as an outgrowth of this untrammelled greed for profit, a competition to conquer as many markets and people as possible against any and all rivals. In present times, capitalism remains no less rapacious or colonising. Yet its focus goes far beyond merely dominating and shaping an existing populated world. Now it seeks to establish and spread profitable realities that seamlessly combine the physical and the virtual fuelled by processes and cultures of quantification. Whether it is crunching sophisticated data to create effective (and commonly almost subliminal) digital marketing strategies, tracking one's time for constantly producing a more efficient working schedule, spending money to advance one's 'character' in the latest virtual role playing game or even creatively imagining how a presently depressed building space could be repurposed to become a profitable enterprise – the possibilities and opportunities for manufacturing market realities are currently seemingly endless.

This chapter explores, therefore, the proliferation of monitored and accountable marketised worlds. Building on the insights of Chapter 3, it highlights how present experiences of fluidity create the conditions and means for greater quantification and as such monitoring. Notably, it fosters a desire to account for the shifting dimensions of our environment – pinning down through analytics its unfolding time and space.

Importantly, such monitoring is simultaneously expansive and restrictive inasmuch as it encourages the discovery of ever new realities whose binding feature is their fidelity to market demands regarding profitability. This fosters, in turn, a pervasive expectation that subjects work constantly towards employing big data to make their worlds more valuable. Further, it inundates society with an entrepreneurial ethos to deploy this virtual power to manufacture fresh digital and physical contexts to exploit financially.

Hence, big data shifts conventional colonialism to conquering technologically manufactured capitalist worlds, showing how any and all manifestations of space and time are ripe for exploitation and domination. In turn, an emerging fantasy of the 'fully monitored reality' is produced – the ideal ability to shape time and space to their own personal advantage rather than being determined and colonised by the marketable desires of others. Critically, this growing culture of monitoring masks how the free market system and the financial elites who most benefit from it are increasingly unbound by the constraints of time, space or any social regulation.

Accounting for a Mobile World

The world, it is constantly intoned, is undergoing a rapid change – the likes of which are almost unprecedented. Technology is transforming social relations, connecting people in ways never before even imagined. It is breaking down borders of communication, and in doing so producing new geographies of interactions. It has enlarged the scope of how we talk to, gather information from and even act in concert with other people. It has also created new digital spaces, uniting the virtual and physical into vibrant spheres for cultural exchange and creation. Tellingly, this enhanced fluidity is matched in intensity with a greater ability to quantify these digitised realities.

While technological advances have certainly brought with them fresh social networks, they have also ushered in what can be termed a general ungrounding of what appeared until recently to be stable cultural and physical realities. Previously entrenched communities and populations now find themselves no longer so cohesive or certain in their existence. More precisely, while it is understood that empires can rise and fall, and civilisations come and go over time, there is a general expectation that where one lives will remain relatively consistent, at least within their lifetime. Of course, histories of mass migration – some chosen to an extent freely and others forced upon their subjects – reveal how dynamic a single generation of existence can be. Nevertheless, the goal of such immigration was to establish a new stable beginning, to re-establish home, to assimilate into a secure world that one could find their own safe place within. The new millennium, by contrast, has shifted the very ground from under our feet – both literally and figuratively. The intrusion of virtuality and smart technology into almost all spheres of life raises questions of whether there is even a 'there' anymore. As one popular article in *Forbes* recently declared, 'With all the powerful social technologies at our fingertips, we are more connected – and potentially more *disconnected* – than ever before.'[1]

These changes have led to a general 'reimagining' of community. Returning to the insights of Benedict Anderson and the imagined community – the structural development of the modern state was coupled with a patriotic discourse that led people to have a sense of cultural and political unity with those that they have never and likely will never meet.[2] It is an imagining that would bring millions together in a common identification. Thus, to an extent mass identity has always had a strong virtual component – and one that relied on technological advancements (in this case print publishing) for its proliferation. Physical developments, in this regard, combined with imaginative discourses of belonging to create new and vibrant cultural selves.

However, this digital turn does indeed represent something novel – it is the ability for people to imagine communities and forge identities in cyberspace. Further, it personalises these ultimately virtual associations, granting individuals greater power to create their own networks across physical borders and spaces. In the words of the scholar Keith Hampton: 'Social media has made every relationship persistent and pervasive. We no longer lose social ties over our lives; we have Facebook friends forever. The constant feed of status updates and digital photos from our online social circles is the modern front porch.'[3]

This process of 'reimagining communities', though, does not eliminate anxieties regarding the stability of the time and space of capitalism. For all its empowering of people to log on and create their own networks, the homogenisation of society linked to globalisation has ironically fostered the feeling that there is little room left for places to be culturally unique in the world. Put differently, there are fears that we are headed to a future corporate reality where everything looks the same, populated by identical pre-packaged chain stores, restaurants and housing. Sociologist George Ritzer warns that we are currently living in an era of the 'McDonaldization of Society', characterised by the global spread of corporate culture. Crucial to this process is what he refers to as the 'nothing-something' dynamic of place, whereby corporations represent a 'nothing' place conceived of as a 'social form that is generally centrally conceived, controlled and completely devoid of distinctive substantive content'.[4]

Interestingly, quantification technology has risen in almost exact parallel to this deeper sense of unease. Big data has allowed companies, governments and even individuals to more fully monitor, analyse and make sense of their daily lives and preferences. Wearable technology permits one to keep track of everyday activities as well as deeper bodily processes (such as heart rate and even insulin levels). At a broader level, the internet has made information about social spaces much more widely available. You can now use Google maps to look at almost any place in the world. People can watch uploaded events happening around the globe on Facebook Live in real time. Hence, if the world is becoming much less grounded, it is certainly also becoming increasingly easier to surveil and quantify.

This seeming contradiction points to the emergence of what can be referred to as a 'mobile world'. It is one where space and time are not necessarily stable but accessible and transportable through smart technology. Where people carry their networks and communities with them in their pockets;[5] in which individuals can learn about any place, anywhere and anytime through a quick internet search. Reflected is the prospective emergence of 'mobile time' changing the very 'rhythms' of our everyday lives. Through this emerging notion of mobility, accounting and fluidity merge into a dynamic means for securely navigating and making sense of an often confusing present-day capitalist reality.

In this mobile world, being connected is of the utmost importance. Here traditional notions of mobility are rebooted. It is much more than being on the move. It is about gaining access to ever newer digital networks and information. This is a 'linked up' culture in which one finds their grounding in fluidly 'discovering' new places and people to connect with. Where a sense of ontological security is gained not through a single shared identity tied to any one place or belief but rather the ability to verify, quantify and account for our multiple 'lived-in' physical and virtual environments.

Monitoring Capitalist Realities

Reality in the present era appears to be profoundly divided. Traditional notions of time and space are being continually uprooted and displaced, undergoing a constant stream of updating. Simultaneously, the capacity to account for and quantify these shifting social dimensions is at an all-time high. Emerging from this tension is a novel form of social belonging – one built on entering into dynamic mobile networks empowered by a sophisticated technological culture of digital data collection and information sharing.

Critically, this echoes ideas put forward by those working on actor–network theory (ANT). ANT proposes that humans and non-humans are constituted and exist within evolving socially constructed networks.[6] Particularly relevant to this analysis, is how it attributes agency to technologies as well as human subjects. It reveals how these historically configured networks grant and evolve from the different social affordances and capabilities of the diverse actors that compose it. Accordingly, it must be acknowledged that 'Technologies are not given. Instead they are discursive moves in a never ending cacophony of efforts at social ordering.'[7] Significantly, ANT highlights the cultural basis for reality; or more precisely, how its dimensions of time and space are formed within a broader set of socio-historical relations. This is not to say that they are purely subjective. Rather it is to point to the complex and even conflictual ways different networks produce diverse experiences of temporality and spatiality, ones that cannot be easily separated, or necessarily at all, from these socially manufactured and emergent contexts. Reality is, in this sense, always a contingent social accomplishment – and one that could be otherwise.

Yet where ANT can still be developed, and in ways that are particularly relevant not only for this analysis but also shedding light on the contemporary period generally, is how the cultural discourse of networks impacts and shapes these underlying networked relations. It is precisely this concern that is fundamental to understanding the rise of an accounted-for mobile society. Indeed, subjects increasingly perceive themselves as being active parts of networked communities. They critically 'imagine' themselves as dynamic elements of these embedded relationships, crafting their identity and existence around the ability to move between diverse networked realities.

This then focuses contemporary empowerment and agency on questions of how individuals can take advantage of these various networks. To do so means being better able to quantify what they are and how they can be best accessed by individual users. It is the shift, to this effect, from 'actor network' to 'networked actors' – as opportunity and possibility are explicitly framed within being able to navigate digital configurations often involving a range of connected human and non-human users. Accordingly, information technology

> does more than just change the cost of transportation and communication: it alters the manner in which economic value is created, changes how international production is organized, and reopens basic economic bargains struck around individual liberty and economic rights. There is no inevitable political path driven by technology; rather evolving technology shakes up the political order, creating the foundation for fundamental rights over the organization of markets and politics.[8]

Quantification and monitoring, nevertheless, are the conditions of possibility for this emergent type of network-based fluidity and political transformation. To be fully mobile means having the information necessary to be flexible and adaptive. The more data that one can collect, the more they can clarify, analyse and ultimately assess how one should enter into and interface with these differing networks. Without this knowledge and these techniques of information gathering such a mobile existence would be nearly impossible.

Consequently, processes of big data, internet searches and other forms of digital quantification must be seen as a distinct type of social technology. They provide a cultural framework through which to exist within a networked reality. Yet while these networks are socially dominant, they do not conform to traditional notions of society as such. Put differently, they are culturally connecting but not necessarily hegemonic or singular in their experience. Rather, they serve as sets of embedded and evolving social relationships that help individuals define, design and partake in different experiences of space and time. Accordingly, they are better described as coexisting mobile realities. This echoes Jameson's prophetic description of the contemporary world as 'the fragmentation of time into a series of perpetual presents'.[9]

Nevertheless, this world of shifting realities is one that enhances rather than diminishes the importance of quantification. This new mobile monitoring locates one at all times, making public where people are and what they are doing. Apps such as Foursquare announce to the world where you are presently. The ability to follow people's movements in real time is now a normal part of contemporary existence. Emerging is a 'checking in' culture where we can almost seamlessly and voyeuristically slip in and out of each other's lives as well as different spatialised 'realities'. At play is a type of 'digital tourism' writ large.[10] It also gives users pinpoint accuracy to literally and figuratively navigate any place they find themselves. Programs like Google Apps seemingly make it almost impossible to ever get totally lost in this fluid world. This extends to time, as time stamps on our emails, texts and calls tells those we are communicating exactly when we interacted.

In this respect, it is through such quantification that the flux of the current period is turned into a manageable mobility. Here the mobile smartphone that is so central to our daily activity represents the deeper ability to use quantifying technology to plot a course through a networked life. It anchors us to these realities – telling us in ever greater detail where we are, how long we have been there and how best to spend our time when there. It sets the coordinates for these socially constructed worlds that combine the physical and cyberspace. As the once stable dimensions of our past realities crumble, quantification has, in this sense, emerged to once more make 'real' our present, more mobile ones.

Smartly Managing Your Realities

A defining feature of the twenty-first century is the enhanced ability of people to enter into a multitude of realities. Time and space are now not confined to the physical world. Rather, they are dramatically expanded in their possible expressions through digital networks and virtual reality. However, at the core of these enhanced possibilities is an enhanced demand for quantification. This is matched in intensity by a 'smart' ethos for engaging with these diverse worlds – one that compels individuals to access and manage these realities effectively.

This has led to the emergence of an increasingly accounted-for post-modern reality. The access to a digitally networked society is one that deconstructs and to an extent defies a coherent narrative or any singular way of being. Existing is a 'hybrid model of circulation, where a mix of top-down and bottom-up forces determine how material is shared across and among cultures in far more participatory (and messier) ways'.[11] It is characterised by spatial fluidity and temporal flux. To this effect, 'Cyberspace is a place. People live there', according to Lessig, 'They experience all sorts of things that they experience in real space. For some they experience more.'[12] Existence is fragmented into a diverse set of 'cyber' and physical worlds. Crucially, though, this more dispersed post-modern reality should not be confused with one that is either nonsensical or incoherent. Instead it is permeated by an accounting ethos – as new technologies allow individuals to quantify and navigate these networks more successfully.

Critically, such insights draw inspiration from Lefebvre's groundbreaking social reimagining of 'space'. He famously declares 'space is a (social) product [...] the space thus produced also serves as a tool of thought and of action [...] in addition to being a means of production it is also a means of control, and hence of domination, of power'.[13] At stake in his highly influential perspective is the concept of spatialisation depicting the complex social production of space. These spatial productions combine everyday practices, existing representations of space and the shared 'spatial imaginary' of the era.[14] Fundamentally, what is being produced is not just space but social reality itself, each of which contains its own physical and social rhythms.[15]

This spatialisation, though, has been profoundly digitally augmented. Virtual reality and social media have transformed the social production of space – moving it from the almost purely physical to one that integrates and is progressively dominated by cyberspace. This reflects the rise of what Cohen refers to as 'network space', whereby this digitised reality 'expresses an experienced spatiality mediated by embodied human cognition. Cyberspace, in this sense, is relative, mutable, and constituted via the interactions among practice, conceptualization and representation'.[16] This evolution represents a novel process of what can be referred to as virtualisation. Significantly, this encompasses how different operating platforms, websites, digital networks and physical places are socially reproduced as distinct cultural spaces. The ethos here is less immediately domination and hegemony as it is access and malleability.

Such virtualisation has brought with it, in turn, a novel dynamic for the production of social space. It is one that revolves around the need for ever greater quantification. Conventionally, spatialisation focuses on the stabilisation of time and space – the pinning down of a coherent and stable cultural reality. Virtualisation has profoundly rebooted this process. While it still centres upon the social manufacturing of space through cultural knowledge, it now emphasises the importance of collecting as much data as possible about these spaces in order to discover fresh ways they can be engaged with, accessed and used. The more data available, in this sense, the greater the possibilities.

Space, in this respect, represents an immersive world that individuals can log into and experience, increasingly on their own terms. To this effect, people use digital technology to learn more and discover new things about a given place and the things that populate it. Samuels thus argues that we have entered into

> a new cultural period of automodernity, and a key to this cultural epoch is the combination of technological automation and human autonomy. Thus, instead of seeing individual freedom and mechanical predetermination as opposing social forces, digital youth turn to automation in order to express their autonomy, and this bringing together of former opposites results in a radical restructuring of traditional and modern intellectual paradigms.[17]

On any given street they can see how well a restaurant has been rated by others, if a local business is hiring and when a nearby movie is playing. Likewise, people can share this space through digital technology, introducing it and what they are doing as part of their broader social networks. This reflects an updated version of what sociologist Roland Munroe refers to as 'extension' – describing how individuals use different social artefacts and technologies to 'extend' into a given social reality.[18] Extension has morphed into immersion, as people submerge themselves into a diverse set of socialised spaces, employing quantification techniques both to gain greater knowledge of them and temporarily habituating them in accordance with their personalised desires.

However, this expansion of spatial possibility is itself regulated by a new ethos of properly accounting for time and space. The technology and artefacts – such as the internet and social media – that allow people to immerse themselves in these worlds, also guide them to use these spaces 'intelligently'. Here 'smart technology' not only gives people the opportunity to explore the potentiality of space but just as significantly gives them the opportunity to 'smartly' inhabit these realities. This ethos is witnessed in the rise of 'life hackers' who have found ingenious ways to maximise their daily existence. One example, chronicled in a *New York Times* article, was a technology writer who developed 'a program that, whenever he's surfing the Web, pops up a message every 10 minutes demanding to know whether he's procrastinating'.[19] It is imperative for all of us to 'upgrade our lives' through these hacking techniques.[20]

Fundamental then to virtualisation is a pronounced ethics of spatial accounting, one that demands that people continually quantify how and in what ways they are inhabiting their diverse encounters with time and space. Indeed, one can never learn enough about a space and what is inside it. Every building has a history, every tree in a park can be identified and every shop explored first online. Yet with this new knowledge comes a renewed expectation to properly discern which piece of information is relevant. We must take advantage of 'mapping hacks' that permit us to optimise the new era of 'electronic cartography' where almost everything and every place has been digitally mapped out. Even more so, we are expected to embrace the infinite potential to more 'intelligently' inhabit these environments. People no longer have the excuse to blindly use the space around them – instead they must collect all data available to maximise their utility from them. They have a personal responsibility to manage their realities smartly.

Capitalising on Time and Space

The possibilities for experiencing reality have arguably never been so great. Virtual technology has made previous limits to time and space close to being a thing of the past. Soon if one wants to travel half way across the world, they will simply need to put on a headset, turn a switch and open their eyes to a virtual reality come to life. Even today, people live in multiple worlds, from immersive first-player video games to the different social networks that they access and 'live' within. Yet at a time when the possible uses of space are so high, it is almost wholly contained within the quite narrow ideological borders of capitalism. The ultimate purpose of reality, in all its growing forms, is to be profitable for both oneself and others.

Reflected is the much discussed 'neoliberalisation' of space; more precisely, the turning of all spaces into an opportunity for private gain, the 'rolling out' of marketisation and privatisation across the whole of societal relations.[21] Tellingly, this does not imply the complete uniformity of all time and space. The image of 'modern times', where everything resembles a factory, is a far from accurate picture of the contemporary period. Instead it is composed of a diverse range of realities whose common bond is their ability to be marketised and exploited for a profit. This reflects, in turn, the 'two faces' of science and capitalism in the twenty-first century: 'on the one hand, an economy largely characterised by mundane technologies and globalisation, and on the other a scientific commons continually appropriated and harvested by capital and caught up in political economies of promise'.[22] To paraphrase Mao, 'let a thousand realities bloom, each profitable in their own beautiful way'.

The current expansion of social space has thus produced in its wake an enlarged capitalist demand for profit. It has increased the very scope of exploitation. Space is now less a physical place as it is a dynamic market opportunity. It forms a defining part of an emerging 'cell phone culture'. Hence,

> The much discredited, yet hydra headed notion is very much alive here, as we have seen in the 'good' power to increase dramatically our productivity and social capital, become our life recorder, or help us organize a rally. The flip size of this is the belief that mobile technologies are powerfully bad, inciting us to riot, affray, excessive sociability or solipsism, or crimes against grammar or cultural values.[23]

For this reason, it is imperative that individuals ensure that the diverse and evolving dimensions of social reality must be effectively and smartly mined for all that they are worth.

Reality therefore must be more than just quantified and accounted for. It also must be made financially accountable. Space can be used in a wide range of ways, as long as it is fiscally viable. Sustainability, hence, takes on a rather new definition. It concerns the ability of a place to sustain itself economically. This seeming contradiction is witnessed in the discourses surrounding 'smart cities' that while often promoting themes of greater democracy and empowerment, are commonly really simply 'marketing language for city "potentials"'.[24] One can hypothetically make any reality they want so long as it is profitable. The only limit to the post-modern is the bottom line.

Ever-present in all this is the new 'habitus' of capitalism. The French sociologist Pierre Bourdieu introduced this term as 'dispositions that are both shaped by past events and structures, and that shape current practices and structures and also, importantly, that condition our very perceptions of these'.[25] It is the accumulation of one's life experience in such a way that they come to physically embody social capital. In this sense, the physical and social interact and are dynamically unified as the basis for capitalist reproduction. However, virtualisation adds a distinct wrinkle to such processes. Specifically, it focuses individuals on deploying accounting technology to ensure that they are maximising the value of all their networked realities.

It is crucial that people successfully accumulate and deploy their virtual capital to prosper in these digitised times. In particular, it demands that they assiduously keep track of their time to make sure of its overall value. This speaks to the much-lamented rise of an 'always open' capitalism. As Crary observes, '24/7 markets and a global infrastructure for continuous work and consumption have been in place for some time but now a human subject is in the making to coincide with these more intensely'.[26] It produces, furthermore, an 'empowerment/enslavement paradox' associated with mobile technology in which people 'feared that they had become slaves to the machine'.[27] This has predictably extended to the workplace: while many professionals liked the increased flexibility provided by these technologies, nevertheless 'the same tools that empowered them in their jobs in so many ways also took away long-cherished freedoms in others. Besides "less personal time", study participants frequently cited increased work pressure, closer monitoring and supervision, and the inability to separate and keep distance from work.'[28]

Although these critiques are certainly welcome, and indeed troubling, they only partially reflect how much capitalism has taken over our times. Rather, it is that any and all temporalities, fast or slow, long or short, can now be optimised to achieve full productivity and efficiency. Above all, it is 'an attempt to shape temporal orientations in a more entrepreneurial form'.[29] Quantification, accordingly, allows individuals and organisations to account for their time and therefore ensure that they are always temporally accountable to the needs of capital.

Similarly, individuals are expected to constantly assess how they can make the best use of their space. They are called upon to create a proper 'habitus' that is conducive to maximising their efficiency and productivity. The need to do so is even more imperative given the rise of what is commonly referred to as a 'boundaryless career'.[30] Considering that one can now work seemingly anywhere, the world becomes a mobile office. The goal, importantly, is not to homogenise all realities into one uniform office space. Rather, it is for each individual to ascertain accurately how they can turn places into work spaces that are most suited to their specific professional and market needs.

Capitalism has, hence, truly begun to virtually spread 'anytime and anywhere'. Neoliberalism has made capitalism flexible for a post-modern existence that is as spatially expansive as it is ideologically limited. The more digital technology has allowed people to expand their experience of space and time, the greater the opportunities for the market to adapt and enlarge the scope of its operation. There is now, thus, but one capitalist world composed of many marketable realities.

Virtual Colonisation

A dominant critique of neoliberalism is that it spreads into all spheres of social and personal existence. Its values of marketisation and privatisation are not confined to the economic sphere. Rather, they are universal principles for guiding every and all cultural relations. In the contemporary period, these capitalist ideals have spread even further, using digital accounting and virtual capital to create and discover ever newer social worlds to exploit economically.

As mentioned previously Marx famously compares capitalists to 'vampires', evocatively proclaiming 'Capital is dead labour, which, vampire-like, lives only by sucking living labour, and lives the more, the more labour it sucks'.[31] Colonialism was, in part, a natural outgrowth of this seemingly unquenchable thirst for profit. Again quoting directly from Marx: 'the veiled slavery of the wage-labourers in Europe needed the unqualified slavery of the New World as its pedestal. If money "comes into the world with congenital bloodstain on one cheek" "capital comes dripping from head to toe, from every pore, with blood and dirt"'.[32]

In structural terms, to simply survive companies and states always have to find new markets to conquer. This ethos has extended far beyond the well-chronicled 'Age of Empire'.[33] Instead, it has extended to current processes of neocolonialism linked to corporate globalisation. Looking even further ahead, it is what largely continues to drive virtualisation. Indeed, the conquest of the new age is the use of data to optimise one's use of space.

Accounting, in this regard, should be viewed as a present-day colonising activity. The link between colonialism and quantification has always been strong. In the current era, the emphasis is on collecting all data and information available to determine the best ways it can be exploited. This dynamic is witnessed in the rise of the 'urban entrepreneurialism' associated with 'corporate smart cities' that override and displace 'participatory and citizen-based types of smart initiatives'.[34] The colonising aspect is twofold here: first, virtual monitoring has become a universal feature of all facets of contemporary life; second, it is used as a tool for marketising and ultimately profiting from these spaces, wherever they may be and however they may be accessed.

Consequently, here we see a crucial paradox relating to how neoliberal reality connects to its reliance on virtual power. While spatial possibilities may appear to be ever more infinite, there is decreasingly little room for undefined spaces. The potential for liminality is progressively diminished, as all places must be quantified, monitored and made fiscally accountable. This is reflected, for instance, in the current neoliberal development of 'desakota' places (areas near cities that combine urban and rural elements) – as these ambiguous spaces are filled in places such as the Philippines with profitable suburban gated communities at the expense of poorer farming communities.[35] Foucault refers, in this sense, to the importance of heterotopia, places that are 'capable of juxtaposing in a single real place several spaces, several sites that are in themselves incompatible'.[36] Yet in this post-modern and increasingly quantified world there is no room for in-between spaces.

Contemporary reality is therefore exceedingly flexible and extraordinarily limited. The space of neoliberalism is connected by a shared ethos and demand to discover the possibility of profitability. It is universal in its economic purpose but context-specific in the expression of these desires. This insight echoes Aihwa Ong's seminal reimagining of neoliberalism as a 'mobile technology'. She declares that:

> the very conditions associated with the neoliberal – extreme dynamism, mobility of practice, responsiveness to contingencies and strategic entanglements with politics – require a nuanced approach, not the blunt instrument of broad categories and predetermined elements and outcomes ... Neoliberalism is conceptualized not as a fixed set of attributes with predetermined outcomes, but as a logic of governing that migrates and is selectively taken up in diverse political contexts.[37]

Hence, while the ideological potential for space is rather narrow, the possible forms neoliberalism can take has only grown exponentially. Drawing on this concept, Lombardi and Vanolo tellingly describe how. 'as a consequence of neo-liberalism and economic crisis, local governments are more and more in charge of providing urban services, while the smart city paradigm is offering new areas of economic profitability for private companies promoting technological solutions'.[38]

The present colonisation of space thus rests not in homogenisation; instead, it thrives on its adaptability and creativity. The goal is not to take over a space in the most traditional sense of occupation and rule. It is to, by contrast, exploit any and all spatial possibilities. Further, it is to pre-emptively guide all such potentialities in the direction of being profitable. Emerging is a novel form of creative capitalism that updates the traditional relation of accounting and colonisation. The social theorists Boltanski and Chiapello have gained well-deserved renown for their description of the 'new spirit of capitalism', whereby creativity itself is inscribed with and directed towards profitable ends and the reproduction of the capitalist system.[39] Virtualisation has taken this capitalist co-optation of creativity one step further. Now it is about finding novel and innovative ways to use space. What is crucial, in this respect, is to always be creating new profitable realities.

In the present age, colonialism has taken a new spatial turn. To dominate the given world is no longer enough. Instead, accounting technology must be drawn upon to find ever new ways to profit from existing places. Indeed, places are now sites for creative exploitation. This extends beyond the realm of the physical. It encompasses virtual realities and cyberspace. Through accounting power the present goal of neoliberalism is to conquer all our socially produced worlds.

Never Missing Out on Reality

A critical feature of the contemporary period is the colonisation of social spaces through accounting power. Significantly, colonisation is never merely external in its effects. It also profoundly invades and shapes one's internal sense of self and the world. It is not surprising, therefore, that virtualisation has colonised present-day subjectivity. In particular, attaching it to a new cultural fantasy of exploiting 'always quantifiable' realities.

Critical for understanding the appeal of this cultural fantasy is the intimate relation between psychic security and the social production of reality. Returning again to the insights offered by Lacan, the very notion of a coherent 'reality' itself is a cultural construct created and clung to in order to avoid the 'real' of human existence – one that is fragmentary and always perilously close to psychic disintegration. According to Žižek:

> The ontological scandal of the notion of fantasy resides in the fact that it subverts the standard opposition of 'subjective' and 'objective'. Of course, fantasy is, by definition, not 'objective' (in the naïve sense of 'existing independently of the subject's perception'). However, it is also not 'subjective' (in the sense of being reducible to the subject's consciously experienced intuitions). Rather, fantasy belongs to the 'bizarre category of the objectively subjective' – the way things actually objectively seem to you even if they don't seem that way to you.[40]

The culture of virtualisation only exacerbates this deep-seated anxiety. It reveals just how transitory and fluid these realities can be. Quantification, hence, is a consistent and eternally updating antidote to these psychic concerns. It gives these spaces a stable 'reality' – an appearance of being sensible and coherent.

However, it also brought with it fresh insecurities that threatened to upend this always precarious ontological stability. There is a distinct fear that one is overlooking a potentially valuable use of these realities as well as feeling that they are constantly at risk of being left behind by constantly updating virtual worlds. These fears are captured in the phenomenon of the 'Fear of

Missing Out', representing 'a pervasive apprehension that others might be having rewarding experiences from which one is absent', and marked by 'a desire to stay continually connected with what others are doing'.[41] Taking this anxiety to its logical psychic extreme, if the possibilities of space are relatively infinite and the speed of change is now close to instantaneous, then one is always living in a reality that has just passed them by.

Nevertheless, it is precisely these techniques of quantification that continually act to temporarily keep these existential fears at bay. The ability to constantly collect data about a space gives it a constant (post-)modern 'reality'. A city block does not seem so forbidding – a contemporary concrete jungle – when one can see what it looks like on Google Earth before even arriving. The daunting task of choosing where to eat when there is suddenly so much choice is partially alleviated by being able to draw on apps and the internet to find the place that best suits your tastes. Every place is definable according to your preference and needs. So too can its pace and rhythms be manipulated and managed to serve your own interests. If you are looking for a leisurely stroll you can easily plot the best and most scenic course, or if you are in a hurry you can look on websites telling you the quickest way to get to where you are going.

Reflected is the rebooting of colonial desires for the contemporary digital age. Colonialism was always in part based on a cultural fantasy of controlling others – a desire perpetuated to lessen the insecurity associated with a deeper lack of self-determination and social agency. This speaks to the Lacanian notion of the ego as 'extimate', representing 'an internal externality' that ultimately reflects the culturally imposed 'desire of the other'. The struggle, in the current era, has shifted to using technology to personally take advantage and create space for one's own desired specifications. In this respect, neoliberalism internally colonises the present-day self precisely through making people virtual colonisers, exploiters of ever newer emerging worlds linked to their increased capacity for quantification.

The drive then of the neoliberal colonial subject is the affective longing to 'never miss out on reality'. It is a fantasy that there are ever newer worlds to discover and personally profit from. The overwhelming deluge of information is turned into an appealing discourse of spatial management and control. And yet it is one that is never-ending – an eternal demand on individuals to be world conquers. The call to be creative virtual capitalists is ever expanding and rapacious. To this end, our current psychic survival rests on constantly accounting for and colonising these fresh social realities.

Smart Realities

The world has undergone a veritable information revolution. It has been radically transformed by the ability to collect, analyse and exploit data. It simultaneously makes present reality continually expansive and totally confined ideologically. Almost anything now is virtually possible, and yet it is seemingly impossible to exist beyond the horizons and practices of capitalism. In this respect, freedom is progressively limited to keeping time and making space in a range of ever newer market worlds.

A common assumption, understandably, is that this emphasis on quantification is primarily or even exclusively technology driven. The increased ability to monitor ourselves and surroundings leads, in turn, to a greater culture of spatial accountability. Further, the proliferation of smart technology has played its part in rendering our social environments both more fluid and more knowable. However, this technocratic explanation risks eliding a key social dynamic, one that is all too commonly ignored or left unarticulated. Notably, it is in no small part a cultural reaction to the underlying acceptance that capitalists and capitalism simply cannot be regulated or kept track of in any substantial way. These are the hidden networks of political and economic oligarchs, the unseen streams of global profit and elite relationships that rule our lives from behind the scenes. This feeling of a system that is beyond our control provides fertile ground to embrace the ability to quantify and regulate our personal experience of space and time. In a free market world where accountability is almost non-existent, there is a perverse pleasure in being able to make our own lives fully accounted for and accountable.

To this effect, in order to approximate some form of control people ultimately come to accept having their realities overdetermined by the values of capitalism and demands of capitalists. At the heart of neoliberalism is a desire to spread the market to every part of society and every possible expression of human existence. However, it does so not through homogenisation but rather through processes of fragmentation and differentiation. In the words of Mitchell:

> Neoliberalism is a triumph of the political imagination. Its achievement is double: While narrowing the window of political debate, it promises from this window a prospect without limits. On the one hand, it frames public discussion in the elliptic language of neoclassical economics. The collective well-being of the nation is depicted only in terms of how it is adjusted in gross to the discipline of monetary and fiscal balance sheets. On the other, neglecting the actual concerns of any concrete local or collective community, neoliberalism encourages the most exuberant dreams of private accumulation – and a chaotic reallocation of collective resources.[42]

Accordingly, there is no accountability for the free market in reality. By contrast, it is preserved as a foundation for the cultural existence of space and time, acting as a condition of possibility for social possibility itself.

The free market is now completely boundaryless in its influence and concrete manifestations. By contrast, the majority of its present-day subjects must constantly create flexible boundaries to accommodate its demands on their existence. Tellingly, this extends to the actual construction of geographic borders, as 'smart border programs' were 'developed after 9/11 as a high-tech solution to competing demands for both heightened border security and ongoing cross-border business movement', in the process revealing 'how a business class civil citizenship has been extended across transnational space at the very same time as economic liberalisation and national securitization have curtailed citizenship for others'.[43] There is an interesting parallel here – one that unfortunately exceeds the scope of this analysis – to the ways in which nations seek to popularly create 'secure' borders to deal with the anxieties of a capitalism without any geographic limits or loyalty. As Bauman presciently notes:

> If the idea of a 'free society' originally stood for the self-determination of a free society cherishing its openness, it now brings to mind the most terrifying experience of a heteronomous, hapless, and vulnerable population confronted with and possibly overwhelmed by forces it neither controls nor fully understands; a population horrified by its own undefendability and obsessed with the tightness of its frontiers and the security of the individuals living inside them – while it is precisely that impermeability of its borders and security of life inside those borders that elude its grasp and seem bound to remain elusive as long as the planet is subjected to solely *negative* globalization.[44]

Fundamentally, this reflects how there is a perverse relationship between the increasing domination of people's time and space and the relative freedom granted to the market, in this regard.

Consequently, contemporary neoliberalism has turned those who it has colonised into its present-day colonisers. Everyone is called upon to explore and exploit existing social spaces – to optimise their personal and economic value. Every space is a new world to potentially discover new means to take advantage.

Space is an opportunity, a virtual and concrete place that we must collect data on, use smartly and profit from. Thus, while the market uses us ever more, it produces a new culture of users. People are rapacious in their desire to find worlds that they can make their own and mine for their resources.

There is a deeply affective component to this contemporary virtual colonisation. Marx theorised the fundamental role that surplus labour played to the reproduction of capitalism. In particular, profit is simply the additional money that capitalists can make from workers after they have made enough to survive. This speaks to Lacan's psychoanalytic of 'surplus-jouissance' (or enjoyment). Here, fantasy provides an enjoyment beyond merely seeking to overcome a psychic lack. Instead, such enjoyment 'takes on a life of their own' and must be psychically spent. This transfers onto the current mobile society as well. To have a space that simply serves our needs is no longer enough. Rather, it must constantly be explored, quantified and invested in so as to maximise its potential value to us. Just as the surplus 'jouissance' is directed towards a fanasmatic 'thing' that can supposedly provide full psychic harmony – so too do people search frantically for the perfect space. It is this desire that drives us towards colonising ever newer virtual worlds through the power of monitoring.

Virtualisation has thus transformed individuals into spatial explorers, monitors, producers and ultimately conquerors. They are tasked with using data and information technology to search for fresh profitable realities. And in doing so, they open themselves to the 'smart' colonialism of a largely unmonitored and unaccountable capitalist system that has fewer and fewer boundaries or restraints in its ability to exploit them.

5
Digital Salvation

Every year at the end of January the world's elite gathers together in the resort town of Davos for the World Economic Forum. Amid serious academic seminars and luxurious dinners, business and political leaders discuss how best to rule the world. Its official mission is 'improving the state of the world by engaging business, political, academic, and other leaders of society to shape global, regional, and industry agendas'. Predictably, the issues discussed are very topical and often quite profound, ranging from 'how to have a good fourth industrial revolution' to human rights and sustainable supply chains.[1] However, in 2015 the discussion took a rather surprising turn – the world's elites were suddenly concerned beyond all else with the state of the population's mental and spiritual well-being. Participants were given classes in mindfulness and even asked to walk more than four miles a day.

The critiques of this well-being agenda are obvious and legitimate. It was a blatant attempt to distract attention from the systematic problems of corporate globalisation – as the issues of rising inequality and chronic economic insecurity were displaced by a renewed emphasis on individual wellness. As the noted critical scholars on the 'Wellness Syndrome', Andre Spicer and Carl Cederstrom, observed: 'When people no longer believe in political transformation, an appealing alternative is individual transformation. When the world cannot be changed for the better, we put all our energies into improving ourselves.'[2] These elites presented a market-friendly vision of a brave new world that could be fixed with meditation, breathing exercises and eating healthier. In the impassioned worlds of Naomi Klein:

> Here is what we need to understand: a hell of a lot of people are in pain. Under neoliberal policies of deregulation, privatisation, austerity and corporate trade, their living standards have declined precipitously ... At the same time, they have witnessed the rise of the Davos class, a hyper-connected network of banking and tech billionaires, elected leaders who are awfully cosy with those interests, and Hollywood celebrities who make the whole thing seem unbearably glamorous. Success is a party to which they were not invited, and they know in their hearts that this rising wealth and power is somehow directly connected to their growing debts and powerlessness.[3]

Beyond the critique, however, lay a strong desire for mass personal empowerment. On the one hand, it represented an implicit admission by those at the top that the free market they once promoted unreservedly was in fact bad for the population's overall health. On the other, they now offered people the possibility of being able to cope with this poisonous social order successfully. In true entrepreneurial form, capitalists had found a way to profit from the cure for the disease they were responsible for creating and spreading. Digital technology is an absolutely crucial and relatively underexplored part of this rising global movement for well-being. While people are being asked to sit silently and eat organic, they are doing so with the aid of mobile apps and interpersonal networks fostered on social media. Living in the present means digitally tracking how balanced and healthy you are being in every moment of every day.

This reflected a new direction for neoliberalism. The free market was becoming deeper and turning inwards. It was seeking to become a force for saving and capitalising on our most intimate desires – giving digitised form to our once mysterious soul. And it was doing so using the most hi-tech methods currently available. More and more people are expected to digitally monitor their spiritual health, well-being and social worth – a form of inner surveillance that makes them morally and ethically accountable for being a holistic, balanced and good present-day market citizen.

New Age Capitalism

Since its inception, capitalism has been infused with a deep-seated religiosity. Early industrialist joined hands with religious leaders to justify their exploitation and profits. Nineteenth-century imperialism went hand in hand, usually quite comfortably, with the need to convert and civilise indigenous populations.[4] Beyond this explicit relation, the spread of capitalism contained an evangelical fervour. The market, private property, entrepreneurship and wage labour supposedly held the keys to individual and collective salvation. Quoting prominent nineteenth-century French socialist Phillip Buchez:

> Consider a population like ours, placed in the most favourable circumstances; possessed of a powerful civilisation; amongst the highest ranking nations in science, the arts and industry. Our task now, I maintain, is to find out how it can happen that within a population such as ours, races may form – not merely one but several races – so miserable, inferior and bastardised that they may be classes below the most inferior savage races, for their inferiority is sometimes beyond cure.[5]

Thus, from its very beginnings capitalism projected itself as much more than a purely economic project – it was a spiritual movement of global proportions that was meant to civilise both the domestic and foreign masses.

On the surface at least, capitalism and religion would appear to be a rather strange and even uncomfortable ideological partnership. The former is ostensibly obsessed with earthly gain while the latter is concerned with spiritual redemption. In practice, religion has always been used as a means for supporting the ruling class and their values. Capitalism and capitalists were no different, in this respect. There is a reason that Marx referred to religion as the 'opiate of the masses', as it asked people to cast their eyes upward to a better world while distracting them from its radical possibilities in the real world. Further, the emerging bourgeoisie class were held up as moral exemplars for a new modern age of commerce.[6]

Tellingly, one of the first and most famous critical attempts to understand capitalist development was steeped in religiosity. Weber's now classic theory of the 'Protestant Ethic' argued that it was the aforementioned Christian morality and culture that instilled the values of thrift and hard work necessary for ensuring shared market-based prosperity.[7] While the historical accuracy of this assessment was and is deeply suspect, it set a religious logic for socially explaining capitalism that still remains relevant to this today. Notably, it attributed shared and personal success to one's overall spiritual worth. In the most modern of times, this idea is witnessed in the attempts of right-wing evangelicals to ideologically marry Christianity and capitalism in its promotion of the 'prosperity gospel'.[8]

However, this relationship is far from historically straightforward or uncomplicated. Indeed, religion has been used consistently as a force for critiquing the excesses of the free market. The old canard that the British Labour tradition was influenced as much by Methodism as it was Marxism contains a great deal of truth and is quite revealing.[9] The radical abolitionists, those willing to take arms against the plantation economy and its exploitation of cattle slavery, were inspired by a deep religious fervour.[10] A century later the civil rights movement was led by a Christian reverend and an insurgent religion, the Nation of Islam.[11] In Latin America, liberation theology drew on Marxism to fuel anti-capitalism and anti-colonial revolutionary movements.[12]

At stake in these parallel histories was a struggle of how and in what way religion was used to make capitalist subjects accountable. For the market evangelicals, religiosity was concerned with ensuring that individuals had the proper capitalist spirit.[13] Indeed, the free market and their adherents have been credibly referred to as cult-like.[14] By contrast, for capitalist heretics it was all about ensuring that the free market received its final and proper divine justice for all its earthly sins. There was a middle ground, of course, that set out the religious conditions upon which the faithful could morally be capitalist – evidenced in such practices as Islamic banking.[15]

In the twenty-first century, capitalism itself has evolved into perhaps our most vibrant and widespread religion. More than simply a secular ideology, it has become a sacred modern belief system. Its proponents are dogmatic with an abiding faith in the saving grace of the free market, regardless of earthly evidence to the contrary. 'From a historical point of view', according to former Nobel Prize-winning economist Joseph Stiglitz, 'for a quarter of [a] century the prevailing religion of the West has been market fundamentalism. I say it is a religion because it was not based on economic science or historical evidence.'[16] To question this faith is to risk being labelled as not only 'irrational' but intentionally or unintentionally contributing to the spread of wickedness in the form of socialism or fascism.

Operating alongside this market fundamentalism was a fresh form of capitalism that merged materialist and traditionally non-materialist pursuits. Emerging from the post-war era was what the famed theorists Boltanski and Chiapello referred to as a 'new spirit of capitalism', which sought to commodify and exploit desires for creativity.[17] Particularly relevant to this analysis was its attempt to co-opt the entirety of the human experience, turning all aspects of people's lives into a labour opportunity.[18] It wasn't merely that people were selling out but rather that they were expected to increasingly explore their inner artistic spirit so that they could mine its economic value.[19]

The advent of the Great Recession, brought on by the 2008 global financial crash, created a new spiritual crisis for capitalism. Notably, it put faith in the free market into question. People were suddenly looking for answers – ones that went beyond accepting morality tales that hard work and good investing would make you rich here and in the hereafter. Instead, the dogmatic foundations of modern capitalism seemed to be on the verge of collapse, without little to replace its corrupted church and its economic priests. It served as a veritable existential crisis representing 'a profound malaise. The existential crisis of the economy we are participating in today rests primarily on a crisis of confidence. People consume less, have a tendency to slow down accumulation and investment ... a symptom of the lack of fundamental confidence in life and in the future.'[20]

Not surprisingly, the response to this spiritual malaise followed a familiar pattern to the past. Conservatives arose promoting austerity and demanding that everyone – from the poor to governments – repent for their free-spending sins. A growing number of the disaffected invested all their hopes in the dreams of CEO political saviours such as Donald Trump.[21]

The arrival of a new hi-tech 'smart' society did little to alleviate this spiritual flux. It appeared to make people and communities less connected than ever, even as it linked them up into ever new and expanding digital networks. The emphasis on data and the relationships forged through social media were increasingly criticised for being dehumanising. Moving beyond the scope of traditional religion, this technology was thought to be cause of our inability to centre ourselves and as an impediment to our well-being and even enlightenment. It appeared to leave us information rich and spiritually poor as human beings.

Into this spiritual abyss, a new ethos of wellness and mindfulness was rapidly arising. Suddenly, there appeared to be a consciousness shift emphasising the need for personal well-being and spiritual nourishment.[22] New age ideas that were once on the fringes were becoming culturally mainstream and socially accepted. The encouragement of 'mindfulness' by business leaders exemplifies this trend:

> The rapid rise and mainstreaming of what was once regarded as the preserve of a 1960s counterculture associated with a rejection of materialist values might seem surprising. But it is no accident that these practices of meditation and mindfulness have become so widespread. Neoliberalism and the associated rise of the 'attention economy' are signs of our consumerist and enterprising times. Corporations and dominant institutions thrive by capturing and directing our time and attention, both of which appear to be in ever-shorter supply.[23]

Just as significantly, these quests for deeper insights and alternative ways of being were becoming big business. We were entering into novel times – 'the new age of capitalism'. Critically, it is one that, as will be shown, asks us to account for and monitor not only for our material value but also increasingly our spiritual worth.

Digitally Grounding Ourselves

Social media and digital technology were heralded as economic and cultural saviours. The information age was meant to liberate our democracy, civic society and economy. It would create new jobs, new ways of communicating and more responsive forms of governance. Yet lost amid the instant messaging, twenty-four-hour news cycle and global networks was a growing spiritual disconnect. The rise of a hi-tech globalisation did not just leave people economically behind, it also left behind a mass cultural and inner void.

On the surface, current 'smart' advances appear to be the antithesis of profound spiritual well-being and transformation.

They evoke a society that is addicted to their mobile phones, for whom meaningful relationships are exchanged for fleeting text-based encounters. This has led, in turn, to a widespread social outcry that such digital technologies are destroying our communities and lessening our meaningful human connections. As previously discussed, the scholar Ben Barber prophesied this deep-seated modern conflict in his groundbreaking work from the early 1990s, 'Jihad vs. McWorld'. This conflict has been 'smartly' rebooted to reflect the supposed divide between a dehumanising technological reality and the desire by a growing many for a more spiritually fulfilling existence.

Underlying this desire for greater depth was a prevailing fear that people themselves were simply no longer needed. The immediate source of this concern was the prospect of automation and robots taking our jobs and in doing so making us economically irrelevant. As the economic editor of the *Guardian* ominously proclaimed as recently as early 2018, 'Robots will take our jobs. We'd better plan now, before it's too late.'[24] Fuelling such worries was the increasing sense that we were being made into efficient free market machines, where 'who we are' matter less than how well we perform. And indeed, neoliberalism to a certain extent is largely subjectless, focused on our lives, hopes and dreams as objects to exploit economically for maximum profit.[25] We were losing not only our identity but our very sense of self entirely.

There is an obvious assumption, therefore, that monitoring technologies would only exacerbate this sense of spiritual loss. There is a long tradition of portraying technology as soulless. Social media similarly is accused of making people into 'soulless creatures'.[26] In the past, the shiny metallic images of industrial progress were viewed as hollow artefacts of an indifferent mechanised modern world. Today this image has been replaced by the belief that our digital lifestyles are artificial – catalysing a return to all things natural and 'organic'. Accounting technology is simply the latest manifestation of this attempt to supposedly turn us all into mere data points, robbing us of our inner humanity.[27]

The cultural emphasis on personal wellness and well-being certainly echoes these desires. It is a longing to assert oneself once more as a unique and important person. To reaffirm that your worth transcends your expected productivity and what an algorithm says about you. It has been referred to as the 'dehumanisation of decision-making'.[28] Even the traditionally conservative and free market boosting *Financial Times* linked this to previous forms of dehumanising management such as Taylor's 'scientific management'.[29] By focusing on well-being you are regrounding yourself in a digital society that feels increasingly socially disconnected and virtual rather than physical. The worry is that, 'From natural disasters to the scale of government spying, we don't seem able to process figures we can't relate to. So will we fall into big data's empathy gap?'[30] The solution is, at least in part, to 'unplug' from technology and reconnect with our concrete selves.

Yet it is also an ironic and rather strange inversion of this neoliberal accounting culture. Rather than collecting data about our efficiency we are suddenly just as concerned with tracking data about our deepest, most essential self. We want to know what fuels our bliss and account for what makes us personally happy. Emerging is a new form of 'cultural analytics' that 'is interested in everything created by everybody. In this, we are approaching culture the way linguists study languages or biologists study life on earth. Ideally, we want to look at every cultural manifestation'.[31] The inner world, in this respect, has become externalised – its own set of data points for identifying and contributing to understanding what was previously considered ineffable and spiritual. It is nothing less, according to a *New Yorker* article, than 'Big Data for the soul'.[32]

The soul is now no longer shrouded in divine or existential mystery. It is discoverable in how you live and more importantly, using the latest big data techniques, how you can live better. The need to 'ground ourselves' in a digitalised world is reversed, subverted into a fresh expectation that you are constantly accounting for your spiritual self. It is a desire for achieving 'online groundedness'[33] in our daily lives. Who we are, in the most profound sense, is now able to be digitally analysed and answered. All that is required for our earthly and spiritual well-being is being more present with our inner data and mindful about monitoring ourselves.

Deeper Data

It is commonly proclaimed that we have entered into the era of big data. It may be more accurate, though, to declare that we are in the period of 'deeper data'. This concept is similar but not identical to machine-learning models of 'deep learning' that seeks to use learning data representations and artificial neural networks in order to enhance predictions and learning. Deeper data refers instead to the mining of our inner psychological and spiritual worlds – the turning of these profoundly personal aspects of ourselves into data resources to collect, commodify and exploit.

On the surface, big data is primarily concerned with observable behaviour and preferences. It tracks what you click on, what you buy, what you post, how you breathe, where you go and even how much you walk. From this large and wide-ranging set of data, it is able to digitally reconstruct you as a person. However, this cyber-identity will always be incomplete. On a subjective level, it reflects the feeling of alienation most exhibit towards their culturally produced social selves.[34] Yet it also poses a challenge for data analysis to dig deeper into who 'we really are'. It is what the sociologist Andrew Abbott referred to as 'an extensive commodification of important parts of previously esoteric knowledge'.[35]

This discovery of our most secret and hidden information opens the digital floodgates to a much more invasive form of data mining. Everything about everyone, at least in theory, could now be discovered. It digitally locates data, 'residing in the deeper emotional layers of the self, the spiritual self reveals itself through one's feelings, intuitions, and experiences'.[36] Achieving wholeness was increasingly and inexorably linked to compiling as much personal data about yourself as possible over the most wide-ranging areas. Even beyond the practices of hi-tech data collection, the culture of wellness progressively revolved around a strict monitoring of one's experiences and responses. This ethos is exemplified in the spirit app SoulPulse, which allows people to track their 'spiritual data' – such as your daily spiritual practices and religious experiences – in real time. Quoting co-founder Jon Ortberg, a senior pastor at Menlo Park Presbyterian Church, 'Your soul matters more than your body. So the ability to monitor your inner, deepest self, your emotional and spiritual well-being, with real-time, realistic information is very valuable.'[37]

There is a growing need for marrying our deepest selves to data. It represents a modern and smart means for engaging with those ineffable parts of who we are and aspire to be. It promises to unlock our cosmic potential. The hope, or for some the fear, is that 'digital technology and neuroscience will combine to create a new understanding of the divine'.[38] Self-tracking is a modern extension of a classic desire to 'know thyself', and an even older longing to transcend it. Thus it is only through big data that we can begin to unravel the mysteries of the cosmos and our soul. As the renowned technology theorist David Berry presciently observes, 'Computational technology has become the very condition of possibility required in order to think about many of the questions raised in the humanities today'.[39]

In the new millennium, this is precisely what allows us to discover our 'inner selves'. To this extent, 'it implies a mode of normalisation that is (1) derived from reality, rather than imposed, (2) relative, rather than absolute, (3) flexible, rather than rigid and (4) plural in scope and scale, rather than individual'.[40] The soul becomes essentialised precisely as a reflection of shifting realities and calculations as the technology becomes invisible, dynamic and totalising.

Just as we expand outward to manufacture and optimise different 'realities', so too do we expand inward as expressions of our 'real selves' become the optimising of our different 'realities' into a totalising soul representing 'who we are'.

What is crucial, in order to mine the true depths of ourselves, is to engage in constant and rigorous forms of 'soul tracking'. This involves moving beyond the mere meditative and exploring the unconscious and neurological forces shaping our existence. It means being truly mindful of our minds – quite literally. At the vanguard of this digital spiritual revolution are Silicon Valley individuals using this technology to guide them on a quest for a sacred and data-based enlightenment. This new age form of 'consciousness hacking' involves 'pulling the neural triggers that can produce the same kind of enlightenment that lifelong meditators experience. Want an out-of-body experience? We have virtual-reality simulations for that. Want to be smarter and happier? You can learn to quiet your pre-frontal cortex – that inner critic – and access more of your brain's attention-focusing norepinephrine.'[41]

What this reveals is the transcendence of big data to deeper data. Its depth is reflected in its almost religious qualities of helping us to discover and nourish our souls. It is, in this respect, a contemporary religious acclamation – declaring our shared faith in data for delivering us from evil and giving us access to the most sacred of qualities.[42] It holds out the promise of 'cybergrace', catching a glimpse of the divine in its algorithms and the surprising personal discoveries provided to us by our data.[43] The deeper we mine our data, therefore, the closer we supposedly get to fully realising the human spirit.

Achieving Data Balance

The infiltration of digital technology into all areas of our lives has led, perhaps rather predictably, to renewed desires to 'go back to nature'. To leave our 'always on' information society behind and return to what is 'true' and 'natural'. The artificiality of our modern reality can only be cured, it would seem, by re-embracing our organic roots, reconnecting physically with each other and rediscovering the concrete non-digital world in which we live. Present-day spirituality, it appears, starts with looking up from our screens long enough to appreciate what is actual and real.

It is interesting that the physical and earthly realm has emerged as a source of purity and increasingly 'the divine'. It was precisely this secular reality that was criticised for so long as being that which kept us away from our most spiritual and deeper selves. Of course, this transcendent notion of an other-worldly God continues to thrive, yet even these traditional forms of worship are intertwined with a renewed focus on the need for spiritual human-based communities.[44] What becomes apparent is how the digital has become denigrated as the source of all that is artificial and non-spiritual. Whereas in previous eras it was society and personal relations that diverted us from the path of spiritual well-being and salvation, now it is our virtual worlds.

The antidote to this virtual problem is to promote digital wellness. This involves teaching people and insisting that they monitor their use of digital technology in order to protect their mental health and physical wellness.[45] There are particular fears that the younger generation is being consumed by social media and mobile technologies, portending a dire future of unsociability and lack of real human connection. It is the increasing responsibility of organisations, governments and parents to ensure the data wellness of their employees, citizens and children, in this respect.

Running parallel to this partial rejection of our digitised present are the growing opportunities to use smart technology to enhance one's spirituality and personal well-being. New apps, easily downloadable, can help track your meditation, provide you prescient tips for being mindful and aid you in living an all-round healthy lifestyle. This is reflected in 'the rise of spiritual tech', as

> 1,000 meditation apps are now available, with big players like Headspace ushering in 16 million downloads as of last year and raising nearly $100 million in investments. Others like Calm, Insight Timer, and Aura are also supporting millions of people through every phase of their meditation journeys. This rise of meditation apps, and app culture in general (as of last month, 2.2 million mobile apps were available for iOS download), has made way from more niche spiritual offerings to appear on our screens too. Golden Thread Tarot lets users 'mirror the digital experience with the physical' by programming their tarot decks into a mobile version that they can call on at any moment. iLuna lets downloaders track the moon through its phases and gives advice on how to live in greater communion with this natural cycle. Most recently came Co-Star, a sleek astrology app that allows users to access their natal charts and compare them with their friends'. When it first launched late last year, it repeatedly crashed due to demand – over 1,000 downloads an hour at times. The app, which uses NASA data to generate personalised daily horoscopes, continues to appeal to millennials craving a new wave of astrology: one that is easily accessible, stylish, and shareable.[46]

Going even further, are fresh ideas that combine cutting-edge 'smart' technologies with conventional spiritual techniques. In a more 'Western' vein, religions such as Judaism, Christianity and Islam have all drawn on the internet to encourage digital spiritual communities. Hence,

> The importance of the web in everyday life – from banking to shopping to socialising – means that religious organisa-

> tions must migrate their churches and temples to virtual real estate in order to stay relevant and to be where the people are. Religious leaders have websites, blogs and Twitter feeds, there are email prayer lines and online confessionals, social networks for yogis and apps that call the faithful to pray.[47]

For this reason, Sister Catherine Wybourne, prioress of the Holy Trinity Monastery in Oxfordshire, and @Digitalnun on Twitter, declares 'Being web-savvy should be a required skill for religious leaders in general.'[48]

Alongside this explicitly spiritual focus, these same technologies are being trumpeted as an ethical force for self-improvement. The internet is a prime source of information about leading healthier lifestyles, augmenting our happiness and achieving our personal goals. We are progressively in the midst of what Cederstrom and Spicer have referred to as a time of 'optimisation'.[49] Big data gives us novel possibilities to realise 'our best selves' and lead the ethical and fulfilling existence that we so passionately long for. It has become a 'distinctive means to monitor and render human behavior and the applications aimed to persuade people to change their practices of everyday life, in areas related to health, mood and fitness but also with few references to sports, training, social networking, transportation, consumptions, emotions and communications'.[50]

These diverse but linked phenomena reveal the dynamic and rich interweaving of digital technology into our innermost desires. Consequently, 'software quite literally conditions our existence, very often outside of the phenomenal field of subjectivity'.[51] It permits us to discover, track and enhance our spiritual and moral behaviours. Critically, these smart advancements are reframed from being that which prevents us from mining our spirituality and building ethical communities, to a necessary tool for making these deeper pursuits a reality. Social media, perversely, exists as a 'site of truth' – where our real selves can be revealed in an offline and online world where 'everybody lies'.[52]

The fact that such data often reinforce biased assumptions, and can be a 'toxic tech',[53] is culturally cleansed by the possibilities of digital spirituality. Yet it also sets the stage for a new form of digital surveillance that wants to monitor and exploit not just your actions but also your very modern-day soul.

Inner Intelligence

Digital technology has been the cause of much spiritual condemnation and moral handwringing. It reflects a disconnected society, whose strongest and most vital relationships are increasingly found in an unreal virtual world. Yet in practice, these networks and mobile technologies have aided and abetted people's contemporary quest for personal fulfilment. It is a digital passport to a better and more profound and ethical existence. It also catalyses and legitimates an insatiable demand for ever more data about one's inner life for this sacred purpose.

Critically, spirituality and ethics become, perhaps above all else, data-gathering exercises – a case of looking inwards by extending digitally outwards. Here it is the use of big data to control 'the soul', denoting 'the creation of human beings who control themselves through self-control and who thus fit neatly into a so-called democratic capitalist society'.[54] Personal wellness and deeper concerns of the soul were suddenly quantifiable data points to be continually collected and analysed. One's daily digital progress was an 'always on' confessional, counting up and displaying our sins for us to check compulsively. The scope for this sacred data gathering is as expansive as compiling information about the moral state of the world and as personal as the monitoring of a person's daily breaths.[55] To this end, spiritual depth was progressively equated with the constant mining of our various 'wellness' indicators.

What we now see sprouting up, as discussed previously, is whole new spiritual data industry. There is a growing number of algorithms and 'smart solutions' for monitoring and improving creativity, work–life balance and wellness. Ostensibly, these all have a quite empowering ambition – to allow individuals to better manage their lives and use the latest technology to explore how they can do their part to make the world a better place, and even inch ever closer to inner peace and spiritual enlightenment. These desires are witnessed in the cultural linkages between big tech and spirituality. Steve Jobs not only helped to build one of the largest corporate empires in the world, he was a dedicated practitioner of Zen Buddhism. He was lauded, in this regard, for being not 'only a pioneer in computer technology. He was also a pioneer in the technology of the brain.'[56] Beyond such 'spiritual' capitalist icons, new products such as the app Headspace are providing people with 'digital therapy' by encouraging them to meditate.[57]

While these digital techniques can certainly provide concrete benefits, they also justify and nourish an increasing surveillance of our 'soul' – the monitoring of our deepest and most profound desires. It is the digital exploration of who we 'really' are in the most cosmic and invasive sense possible.[58] Our modern spirit quests are transforming into a global demand for existential information about ourselves and society at large. It is a constant barrage of looking even deeper into 'who we really are', individually and collectively, and how we can achieve better living through the alchemy of digital discipline.

Central to this rebooted spirituality is tracking not only our actions but our stated and tacit beliefs. Social media has served as a magnificent reservoir for analysing and understanding our preferences and increasingly our fundamental ideas. Algorithms can now track what we believe, our 'ideological blind spots', our vices and how we judge essential notions of moral work and social value.[59] The supposed positive of this process was that it helped to identify and uncover previously invisible prejudices – giving digital voice to conventionally marginalised groups and showing how ingrained biases against them remain. Yet it has also held out the danger of shaping who we are as ethical subjects, and often for quite exploitive purposes.

Indeed, the efforts to put our ideas into practice are intimately linked to this prevailing digital culture. Not only does data open a doorway for truly 'knowing ourselves', they also record how diligently and successfully we live out our beliefs in real life. Our moral and ethical goals are intertwined with the ability to regularly track and assess our daily practices. Social media provides a further platform for fostering ethical networks and realising these common values as part of a wider online community.[60] Our growing ethical surveillance is inexorably attached to our ongoing personal and shared growth. As such, 'The analytics of bodily and mental functions is no longer the privileged domain of professionals ... Everyday analytics progresses with the aid of new devices; however, these are only successful in moving and recruiting consumers if they promote emotional and practical engagements'.[61]

What this uncovers is an insatiable desire for our ethical and spiritual information. We have become cosmic and progressive data subjects. This 'new age capitalism' is driven by an unquenchable demand for our creative, moral and sacred selves. It seeks out fresh areas of our deepest inner recesses in order to discover fresh spiritual resources to capitalise on. The eternal need to find new markets has turned inwards. To survive and thrive we must rely on our 'inner intelligence', the immaterial labour of contemporary capitalism transformed into a spiritual and ethical journey into personal wellness, existential growth and social betterment. This mining of our deeper data serves, in turn, as a portal to our increasingly exploitable digital souls.

Accounting for your Neoliberal Soul

In 2009, self-help guru James Arthur Ray was on top of the world. He was on the fast track to becoming rich and famous for his theories of 'harmonic wealth'.[62] At its heart it was an attempt to combine financial success, spiritual awakening and personal empowerment into an integrative framework for living a fulfilling and prosperous life. Ray preached the gospel of 'harmonic wealth' across the US, attracting a growing and diverse array of paying followers. His methods of self-help and inspiration involved directly challenging audience members and inviting particularly dedicated ones to 'spiritual warrior' retreats where he would test their commitment to his principles by asking them to shave their heads and undergo intense physical ordeals. On 8 October 2009 this self-help movement came to a tragic end as three people died from heat exhaustion in a sweat lodge at one of these boot camps in Arizona. Revealed was the stark danger of trying to save people's neoliberal souls.

What is becoming readily apparent is how the insatiable surveillance of our 'deeper data' is evolving into a means for judging our overall personal, ethical and spiritual self-worth. Wealth, professional success and online followers are indicative of your broader value as an actual and virtual person. Those who have used data and social media to materially and non-materially enrich themselves are held up as contemporary icons for existing in a competitive, networked, information-saturated cyber-world. Inversely, those who are poor are increasingly blamed for their condition based on their inability to 'bridge the digital divide'. More precisely, it is their responsibility to take advantage of these virtual opportunities as well as productively use this technology. Such social judgements have taken on an evangelical character, revealed in the infusing of the 'prosperity gospel' with technological discourses of online wealth creation.

These market-friendly moral and spiritual evaluations support the emergence of a fresh spirit of data-based neoliberal religiosity that unites the ecclesiastical and the capitalist. The mining of our 'deeper data' – turning it into concrete quantifiable evidence – serves as an accounting exercise for uncovering our modern-day market sins. Data analysis is the contemporary confession, our managers and their sacred algorithms are the priests responsible for urging us to be better data believers, while forgiving us for our digital trespasses. What we are witnessing is an updated version of what Foucault refers to as 'pastoral power', modelled after Christian priests and their care and authority over their parishioners. Importantly, 'this form of power cannot be exercised without knowing the inside of people's minds, without exploring their souls, without making them reveal their innermost secrets. It implies a knowledge of the conscience and an ability to direct'.[63] In our times, this pastoral power manifests itself in the continual encouragement to bear our data souls and expose our deepest secrets in order that we may optimise our earthly and spiritual existence.

Absolutely crucial, in this regard, is our capacity to properly and intelligently manage our deeper data. We must learn to appropriately balance our various professional, personal, and spiritual needs through improving our data management skills. This almost explicit embrace of our multiple social identities, the implicit acceptance of our intersectionality, is now used against us in the smartest and most exploitative way possible. Individuals are asked to monitor all of their complexities in order to ensure that they are economically and spiritually worthwhile. Even the most sensitive biological changes are meant to be categorised, tracked and made compatible with this romanticised version of the balanced self at home, at work, and in the higher planes of the cosmos. 'Wearable technology' exists as a type of 'data for life',

> marketed as digital compasses whose continuous tracking capacities and big-data analytics can help consumers navigate the field of everyday choice making and better control how their bites, sips, steps and minutes of sleep add up to affect their health. By offering consumers a way to simultaneously embrace and outsource the task of lifestyle management ... such products at once exemplify and short-circuit cultural ideals for individual responsibility and self-regulation.[64]

Undoubtedly, this cultural connection of data to economic and spiritual agency masks its colonising and exploitive effects. Significantly, 'The algorithm enabled by big data ... stands between the calculating subject and the object calculated; it refracts the subject-centred world. Together algorithms and data filter what we have access to, produce our texts with unseen hands and unknown logics, and reshape our texts, rendering them contingent, mutable and "personalized"'.[65] For this reason, it is worth thinking about big data as data that are captured from us rather than given to us.[66] Indeed, even our use of emojis are collected to gather data about people for market exploitation, thus representing a form of 'affective labour'.[67]

Yet this sense of digital responsibility has a much more existential meaning as well. It is designed to make us look inward in order to avoid having to look outward at the state of the world and our own lack of agency in it. Indeed, 'individuals strategically reveal, disclose and conceal personal information to create connections with others and tend social boundaries'.[68] Just as importantly,

> Such decisions do not occur in a vacuum but as part of an asymmetric power relationship in which individuals are dispossessed of the data they generate in their day-to-day lives ... the asymmetry of this data capture process is a means of capitalist 'accumulation by dispossession' that colonizes and commodifies everyday life in ways previously impossible. Situating the promises of 'big data' within the utopian imaginaries of digital frontierism ... processes of data colonialism are actually unfolding behind these utopic promises.[69]

The passionate longing for inner peace and earthly abundance masks our fundamental inability to shape our own personal or collective destinies. In a present world where so much is wrong and contemptible, unfulfilled and ethically deplorable, the very least we can do is seek to nourish ourselves by any digital means possible. To do so means being willing to mine and monitor the deepest parts of ourselves so that we may be transformed into a more balanced, present, spiritually whole and economically valuable person.

Digital Salvation

In the iconic movie *The Matrix*, after much training the main character Neo, played by Keanu Reeves, achieves complete digital enlightenment. Freeing his mind from the hi-tech-coded illusion that had previously been his only reality, he was now able to see 'the truth' of his existence. He and his fellow humans were being used as 'batteries' for the sentient machines that secretly ruled them. This revelation led him on a heroic quest to control the Matrix and save humanity from their non-human overlords. Made over three decades ago, the film continues to have mass appeal. In addition to its still impressive fight scenes, it speaks to a fear that we are being secretly controlled, that our entire existence is trapped in a simulation not of our making, and that true spiritual enlightenment is found in being able to control this Matrix for ourselves.

The ability to turn inwards and deeply mine the data of our souls serves as the very basis upon which we continually secure our sense of self. Returning to the work of the French psychoanalyst Jacques Lacan, it represents our direct engagement with the 'death drive', the always present fear of complete and total subjective disintegration. Rather than simply ignoring or avoiding it, the psychic constant drive towards death eternally shapes us as social subjects, ironically giving us continual subjective life.[70]

In the present-day case of deeper data, it is our digital attachment to spiritual betterment and personal wellness that maintains our identities. In 'smartly' searching for our best and most essential selves – tracking it daily in how we breath, how we walk, how we eat, how we engage with others, how we deal with our most profound biological life changes – we reinforce our actual existence as a real person, not just a neoliberal economic machine or a set of depersonalised data points on an organisational spreadsheet. It is thus paradoxically through our disembodied data that in the modern age we most fully experience our physical and unique selves. It is more than simply a continual and all-pervasive attempt for self-improvement. Instead this digital monitoring of our spiritual and moral worth is a way to regularly keep track of ourselves in a contemporary hi-tech culture where we are defined by our digital footprint, and what is 'true' and 'actual' changes as fast as our constantly updating real-time Facebook feed.

Central to these rebooted efforts to stave off our inner fear of death is a fantasy of 'digital salvation'. Akin to Neo taking the red pill and leaving the Matrix, through digitally searching and smartly monitoring our ostensibly non-economic and even non-material attributes, we are able to 'see through' our virtual realities and even able to control it. It allows us to slow down time, to make it work for us, to make sense of the buzzing, seemingly infinite amount of data-driven simulation surrounding us and organise it in a way that nourishes who we are and would most desire to be in all areas of our lives. And it is a means to transcend our physical barriers and move to another, almost other-worldly plane of existence where we can tap into our 'purest' selves. Here hacking the system is translated into our hacking our own systems and utterly transforming them in order to maximise our spiritual and economic worth.

These desires for data-based enlightenment are transferred outwards onto the elites who supposedly control our destiny. Suddenly, it is not the economic or political harm caused by these CEOs and politicians that matter. Rather, it is the degree to which they embody strong personal values associated with such personal and spiritual well-being. Their path to riches is rebranded as a spiritual quest to use their entrepreneurial wisdom to achieve a greater good and spread their insights to the world. The corporate executive is transformed into a present-day neoliberal spiritual guru, whose market rationality is now presented as cosmic knowledge that they have tapped into to achieve personal and professional wholeness. They reflect the idealised 'big other' whom we strive to embody through combining technological advances with traditional sacred and existential longings.

At stake is the increasingly universal demand for 'digital salvation'. It is a cultural fantasy that preaches the gospel of smart enlightenment and data-driven wellness. We are saved as unique individuals precisely in our willingness to monitor the well-being of ourselves. The insatiable neoliberal demand for our data has evolved into an eternal personal desire for our 'deeper data'. There is always more information to collect, track and analyse. While we may never be able to completely discover our soul, we can certainly get ever closer through monitoring our most hallowed personal data. For those of a less sacred persuasion, perhaps they can still pursue established humanist goals of secular progress via the gathering of ever more data about their own environmental and social impact. It can be used, moreover, to capture and optimise such ineffable qualities such as creativity and happiness. In opening ourselves to ever greater monitoring we thus fantasise about saving ourselves and the world.

Conclusion: Smart Enlightenment

This chapter has explored the expansion of the present monitoring culture to our innermost selves. There is an increasingly unrelenting expectation that individuals should collect information on their creativity, mental well-being, physical health, moral worth and spiritual growth. This is all done ostensibly in the name of producing greater personal fulfilment and harmony. The digital tracking of our souls is meant to be the pathway in the new millennium to wholeness and prosperity in all spheres of our existence and within all of our selves. Importantly, the aim is not to eradicate these different selves – to eliminate them on the altar of professional success or domestic bliss. Rather, it is to balance all of our various identities from spiritual seeker to hardworking careerist, from doting parent to fun-loving traveller, into an integrated and well-managed digital whole that can be continuously surveilled.

This change can be observed in the evolution from big to deeper data. Experts increasingly highlight the need for 'deeper learning'. It is going beyond more quantity in order to assess the complex factors and causes behind certain behaviours and outcomes. And it is not just an accounting exercise. It is seeking to predict and shape our behaviour and diverse selves. Deeper data is the transformation of this deeper learning into an invasive uncovering of your 'soul'. It mines our previously mysterious and ineffable qualities, the supposed entirety of our internal worlds, to continually monitor and exploit the whole of who we are and could potentially become.

The moral duty of the new age capitalist subject is to explore and exploit this deep data. We have a responsibility to ourselves to ensure that we are well and fit enough so that all our diverse selves can thrive. Personal ethics is translated into a renewed call to 'know thyself' by compiling and tracking all there is to discover about our actions, habits and preferences. Fundamentally, it is to turn this information into a personal, spiritual and economic abundance. It is to reveal to oneself and others what is stopping you from being creative, innovative, social, happy, committed and in touch with the sacred. Data is suddenly life-affirming, necessary for realising our highest self.

Our secular and even other-worldly salvation becomes an exercise in self-monitoring and personal accountability. Left to their own devices, in this respect, are the elites producing the very exploitive system we so desperately need saving from? The misdeeds of CEOs too often escape unnoticed while their philanthropy receives widespread publicity. Ten years after the

Great Recession the capitalist oligarchy has been popularly challenged but still remains firmly in power. Corporations now stress mindfulness and work–life balance while their profound social, political and economic costs are commonly ignored or not followed in real time. It is not that such information is impossible to find or not potentially available – it is that it is of lesser significance to the shared social quest for prosperity, fulfilment and enlightenment.

In the cacophony of publicly available information now saturating our cultural landscape, we can easily become deaf to injustice and the root causes of our collective suffering. To drown out this perpetual white noise we turn inwards, focusing on our own data, our own digital footprint, our own virtual soul in order to cope with the political disconnect from an increasingly disaggregated and disempowering world. In the twenty-first century, data has become the new opiate of the cyber-masses.

6
Planning Your Life at the End of History

Big data and digital technologies are quickly reshaping our world and very sense of self. Yet they are not just reorienting the present, they are also transforming human history. In particular, they are changing how we personally and collectively engage with our past and seek to change our future. Moreover, the very notion of time has been altered, 'datafied', to reflect an individual's own digitally captured experiences and supposedly used to predict their future accurately. We are now in an era where our fate is no longer in the hands of the gods but algorithms.

The effect of this data revolution is already being witnessed in the field of classical history. Specifically, quantitative analytic approaches have been applied to ancient texts and stories for 'achieving greater understanding of our cultural inheritance'.[1] Returning to the present age, people have increasingly become their own self-chroniclers, historians and archivists of their own lives. This is popularly referred to as 'life logging', denoting 'a phenomenon whereby people can digitally record their own daily lives in varying amounts of detail, for a variety of purposes. In a sense it represents a comprehensive "black box" of a human's life activities and may offer the potential to mine or infer knowledge about how we live our lives.'[2]

Yet this hi-tech fascination with our history is also quite paradoxical. It was only three decades ago that the end of the Cold War was meant to signify the broader 'end of history'. This once triumphant proclamation, though, has transformed into a living nightmare of not being able to ideologically reimagine a world plagued by rising inequality, the threat of environmental catastrophe, and social divisions.[3] Rather, it seems, the best we can do is 'recover' a woefully unjust and unsustainable status quo in the face of ongoing economic, cultural and ecological crises.[4] Reflected is a deeper irony of contemporary history in which we are constantly recording our personal transformations in a social reality where failing financial and political institutions remain supposedly timeless and therefore permanent.[5]

Technology, nonetheless, holds the distinct promise of breaking through this historical gridlock. It opens up new possibilities for a 'smart' and more empowering social tomorrow. Indeed, even the thinker who originally declared the 'end of history' now claims that we have entered into a period of 'great disruption'.[6] Such disruptions, however, pose a profound problem for the capitalist status quo and its own history. Namely, how can it continue to present itself and its elites as eternal and unchangeable while portraying people's existence as dynamic, mobile and full of agency? Indeed, the advent of the information age has led to broader existential questions about what constitutes 'life 3.0' as it grapples with the rise of AI and big data.[7] Further, throughout history technological advances have invoked a wide range of economic and social 'anxieties' linked to worries of dehumanisation and morality.[8] These have recently manifested themselves in fears of a 'disruptive' dystopian future where these digital technologies are used against us.[9] If there is any optimism, then it is to be found in being able to somehow find a way to go 'on with the human journey after technology fails us'.[10]

Big data appears to have placed humanity at an existential crossroads. As far back as the 1960s, the economist G. L. S. Shackle introduced what he referred to as 'existentialist economics', whereby

> Decision is paradoxical ... the formal content of these hypotheses, concerning the outcome of each available act, upon which decisions are based, is labelled with dates in the

> future. Yet that future ... has its effective existence only in the present. It is a system of rival figments imagined by the decision-maker in his moment of decision, in his present. The pressures or attractions that bear on his mind are those exercised by imaginations of his, concerned indeed by what he locates at distant parts of his calendar axis which he conceives and uses as a frame of thought, but imaginations are only able to give him experience, or apprehensions of rival and comparable possible experiences, by occupying his mind in the present.[11]

At stake is whether humanity has the ability to reimagine and therefore choose a different present and future society. Or whether they will continue to cling to their 'bad faith in the free market'.[12]

The disruption of smart technologies allows for both possibilities. However, emerging more strongly is their deployment of a self-monitoring culture that promotes the potential for shaping your own personal destiny through big data, while accepting the historical permanency of current neoliberalism. Through predictive algorithms, self-tracking and social media we are now more then ever able to archive our past, quantify our present and forecast our futures. Consequently, humans are data explorers of their own histories, an exploration that provides ever newer opportunities for corporate and government exploitations. To this end, people are increasingly encouraged to smartly 'plan their life at the end of history'.

Running Out of Capitalist Time

Human time is traditionally viewed as being linear and chronological. Put differently, it follows a straight line from the past through the present to the future. Yet, as the renowned physicist Carlo Rovelli presciently reminds us, time is 'something simple and fundamental that flows uniformly, independently from everything else, from the past to the future, measured by clocks and watches. In the course of time, the events of the universe succeed each other in an orderly way: pasts, presents, futures. The past is fixed, the future open ... And yet all of this has turned out to be false.'[13]

This fundamental realisation about time seems to also be manifesting in terms of our current perceptions of capitalist history and technology. Rather than a straight line to progress, where landing on the moon is the forerunner to a globally connected world, it appears we have reached a standstill in which technology moves forward while the majority of humans are left behind. The acclaimed 2013 sci-fi film *Snowpiercer* exemplifies this bleak vision, portraying a future society where environmental catastrophe has forced the whole of humanity to live in a single train that goes along the same tracks endlessly and is divided between elites in the luxurious first class and the rest in overcrowded box cars.

Recent critics of neoliberalism have challenged the progressive narrative traditionally used to legitimise market societies. It has been referred to, in particular, as a 'zombie politics and culture in the age of casino capitalism' – the virulent mixing of an unchanging political imaginary with a high-risk economy.[14] Signified by this other-worldly metaphor is a dead politics that cannibalises any and all attempts to create something radically new. Technology is turning us into post-human 'zombii', consciousless beings that swarm frantically to consume all new innovations without any agency for fundamentally changing anything.[15]

At the same time, history seems to be heading rapidly onward, driven by technology without any human brakes. Theories of accelerationism, from both the left and the right, portend a present-day capitalism that is quickening in speed and making our experience of time ever faster.[16] While there is no slowing down this runaway free market, its velocity can be channelled in either retrogressive or progressive directions. From the right, the speeding up of capitalist production via digital technologies will bring about an intensification of the market and the creation of a more intelligent humanity. From the left, this velocity can be deployed to move beyond the narrow horizons of neoliberalism to reflect the use of technologies for more liberating and revolutionary ends.[17] Tellingly, these innovations can open a window into future possibilities that can then play like a digital feedback loop into the present for making these 'sci-fi' potentials into tomorrow's realities.[18]

However, while the future may be rushing towards us faster than ever, it still seems like a technologically advanced dead end. In particular, this dread is fed by the feeling that there is little humans can do to control our collective fate any longer. The very speed of change appears to have robbed us of our ability to democratically deliberate about where we would like to end up as a society and a species. Instead, we are locked into an inevitable smart 'dystopia' where human politics matters little if at all.[19] To a certain extent this is simply a modern haunting of our colonial histories. The view of technology as an invading historical force is reflective of a recent past marked by the explicit colonisation of one population by another. These fears of the future borne out of the past, are further witnessed in the current cultural preoccupation with invading aliens and their bringing of human extinction.[20] These foretell, in turn, the coming of a more real neoliberal apocalypse, where ecological devastation and environmental decay spell the end not just of our history, but of humanity itself.[21]

Ultimately, it feels as if there is no escape from capitalism, both in the contemporary present and its conceivable futures. It is an era characterised by the 'capitalisation of everything', including our time and aspirations.[22] Concretely, the introduction of social media and digital communication, in combination with corporate globalisation, has created a monstrous 24/7 capitalism that 'never sleeps' and places demands 'always on workers' who can serve it 'anytime, anywhere'.[23]

Things do not get much better when looking ahead to the horizon of a life after work. Indeed, modern capitalism always ironically relied on the individual promise that someday their working existence would come to an end. While socially and ideologically it would brook no possibility of a socialist or anarchist world, on the personal level it motivated people to fully invest themselves in their career through the goal of being able to successfully retire from capitalist labour when they got older.[24] Yet neoliberalism has taken even this promise of elderly escape and respite from people. Our pensions are now insecure and our futures precarious.[25] These fears are exacerbated by the social construction of retirement[26] as being no longer possible – an outdated relic of simpler and more economically secure times. For older people themselves, neoliberalism and its associated technologies have bred renewed anxieties, where established understandings of the self are put into question and their material well-being is suddenly insecure.[27]

Thus the very triumph of the free market that was once celebrated for ending humanity's tragic history and liberating individuals to pursue their dreams, has evolved into a mass resignation that we have no viable future. Big data has contributed to this collective pessimism, disaggregating once coherent life trajectories into easily digestible but often seemingly unconnected byte-sized chunks. As work begins to dominant all aspects of our existence and the prospect of it ever stopping in our lifetime seems less and less likely, it appears that we are finally running out of capitalist time with no place else to go.

Making Histories

If big data has not as of yet radically rebooted our ability to completely alter the course of human history, it certainly has dramatically transformed what it means to be alive. This refreshing of our living existence transcends the usual broadsides that information is rewiring our brains or altering our interpersonal relationships. Instead, it is radically changing matters of life and death. Whereas modern advances in health and technology extended our lifespans, recent discoveries and innovations are threatening to give us a type of digital immortality. As our data are continually collected and saved, our selves can potentially live far past our last breath, potentially recreated through social media and virtual reality.[28] Perhaps in the near future, our data will simply be transplanted into another body using predictive algorithms based on past behaviour to determine our 'reanimated' real-time actions.[29] We may die, but soon our profiles will live on for eternity.

This reconfiguration of the time of our lives has catalysed a renewed need for the monitoring of our digital existence. In particular, this has bred a culture of personally archiving in order to manage our data. Of course, data can never speak for themselves, and as such even before this information-saturated age it was important to think carefully about whom data were speaking for and indeed who was speaking for data.[30] This applies especially to big data, as 'any relationships revealed within the data do not then arise from nowhere and nor do they simply speak for themselves'. It raises serious practical, legal and ethical questions about how such information is collected, interpreted and disseminated.[31]

While these concerns are primarily meant to apply to researchers, they are increasingly relevant to the population as a whole. Data gathering and analysis has gone mainstream, as seemingly everyone is fixated with obtaining information about themselves and others. To this end, digital archiving is an everyday activity for a growing number of people. Such 'iarchiving' highlights 'how recent technological advances both provide new means for self-expression, mobilization and resistance and afford an almost ubiquitous tracking, profiling and, indeed, moulding of emergent subjectivities'.[32] This building of one's identity through archiving requires a huge amount of self-monitoring, as the sheer amount of information and the ability to digitally 'remember it all' can lead to feelings of incoherence and being disaggregated.[33] These play into a broader 'fear of missing out', as the culture of personal data sharing becomes almost instantaneous.[34]

The constant archiving of ourselves is a means for establishing who we are and hope to be. It is a process for building up our personal histories – narrativising our lives – and in doing so making them real and even powerful. This practice reflects theoretically the ways places such as towns or even a 'puny little company in a garage' can become 'centres dominating at a distance many other places'.[35] Similarly, in the contemporary age we use digital technology to mobilise our own 'virtual worlds' to influence others and establish our place in an progressively disaggregated reality. Accounting techniques are integral to this mobilisation as they inscribe people from a distance, thus enabling certain actions while disabling others.[36] The constant accounting for ourselves, therefore, is an opportunity to exert control over our lives. It is part of the broader 'quest to catalogue humanity' in an age where limitless information has ironically made defining and knowing ourselves seemingly close to impossible.[37] In this respect, 'big data is people', as 'The sum of our clickstreams is not an objective measure of who we are, but a personal portrait of our hopes and desires.'[38]

Contemporary self-expression is then channelled into the ongoing managing of our personal digital histories. Through our profiles we chronicle our social existence, putting forward to the world virtual personas that conform to our desired images of ourselves. Social networks provide the space and opportunity to translate ourselves for a cyberspace audience, our offline lives hidden and reframed, linked to our online presentations and performances. Yet the mechanisms for the crafting of 'datafied' selves are themselves inscriptive, shaping and constituting the limits for our identities. Hidden algorithms analyse your histories so that they can not predict but guide your futures. They suggest which news stories and products the 'real you' would be interested in, reinforcing this virtual persona and exploiting it for getting you to maximise your data consumption. It is an updated version of a classic 'sociology of translation', which 'attempts to interrupt all potential competing associations and to construct a system of alliances' so that 'both social and natural entities are shaped and consolidated' in accordance with these understandings.[39] Specifically, it is an instance of 'selves-organising' based on inscriptive performance measures that masquerade as helpful suggestions for living your life.[40]

Significantly, the data-based construction of ourselves is by no means ideologically neutral. Instead, it represents the deployment of discourses and limited allowances from freedom to mask our actual social construction. In particular, it permits us to feel as if we are 'authoring ourselves', both literally and figuratively. Through blogging, texting, Facebook posting and the rest of the ways we interact online we are able to write our own digital histories. It further grants us the opportunity to make sense of the world and who we are in it through such authoring. Yet this creative control of our identities distracts from our ultimate overdetermination by corporations and governments using the latest techniques to redirect our past data to craft our present selves. This echoes the strategic use of 'free choice' by an otherwise regulative capitalist system, as 'the free market must transform "passive individuals coddled by the paternalism of socialism, and characterised by a pessimism, dissimulation, an attraction to populist demagoguery, and a lack of civic virtues" into "a new ethic of the active, choosing, responsible, autonomous individual, obliged to be free, and to live life as if it were an outcome of free choice"'.[41]

Here the traditional 'entrepreneurial self' of neoliberalism is transformed into the 'authoring self' of the information age.[42] More importantly, through writing our own histories in the real-time annals of cyberspace we lose sight of just how much they have already been created for us.

Data Mining the Present

In the new millennium, data has become perhaps the most valued commodity we have. While the desire for oil still starts war, the demand for drugs continues to cause needless global deaths and the need for water promises to be the desperate battleground of the future – it is the requirement for and competition over information that perhaps overrides all else in terms of significance. It is an infinite but precious resource that must be simultaneously protected and shared, withheld and used. It holds the proverbial keys to a better tomorrow today – with the prospect of smarter cities, smarter people and smarter societies on the near horizon. Even more so, it is the very basis for the making of our personal and collective historities. It is not complete hyperbole to claim that in the present day 'you are your data', as 'detailed quantifiable data has become valorised above other forms of information about one's life, health and wellbeing'.[43]

It is not just imperative, in this respect, to discover new, rich veins of data, but also to be able to continually reinterpret existing datasets for producing ever newer insights. People, therefore, must be both data explorers and data miners. Technically, data mining refers to the statistical techniques and machine-learning methods employed to extract patterns and usable insights from raw data. It is the analysis component in the larger process of 'knowledge discovery in databases'.[44] Such mining is strategically employed by firms for a range of different exploitive uses, such as finding meaning in the 'online chatter' of potential customers.[45] They also can contribute to the practices of 'identity mining' and 'identity discovering' that are meant to enhance digital forms of 'identity management' for purposes of strengthening online security.[46]

While data mining represents a complex method for analysing data, it additionally signifies a cultural shift in the ways we understand and engage with our data. Notably, it is a mindset that is reorienting social and personal time, rendering people increasingly as 'real-time' information subjects. Our digital identities are constantly vulnerable to attack and theft and therefore require immediate 'deception detection techniques' for their safeguarding.[47] Customers desire an 'industrial internet of things' so that they can trace in real time the cause of any manufacturing or shipping defects in their products and have it replaced as soon as possible.[48] It further entails the use of efficient algorithms to discover real-time patterns across multiple personal data streams such as health, finance and social media.[49]

The insatiable thirst for instantaneous data analysis has led, in turn, to a veritable data 'arms race' regarding who can deliver this information the fastest. Corporations are constantly promoting and exaggerating to clients and the general public their ability to mine and process data at speeds once thought unthinkable. Importantly, this is as much a social construction as it is a technological possibility. Indeed,

> speedy analytics are central to the spread and intensification of data-led decision-making, governance and ordering processes. The promises of real-time knowing are one means by which organisational speed and agility are seen to be achievable. The result is the pushing back of the limits of datafication ... industry taps into, cultivates and then attempts to deploy the wider rationality of a need for speed.[50]

This veritable 'need for speed' exchanges critical reflection – which requires time and deliberation – for the thrill of immediate and 'objective' knowledge. Moreover, it places our identity in the hands of unseen algorithms and the people who control them, as our self is made and remade in real time through such data mining. Consequently, the 'logic of big data analytics, which promotes an aura of unchallenged objectivity to the algorithmic analysis of quantitative data, preempts individuals' ability to self-define and closes off any opportunity for those inferences to be challenged or resisted'.[51]

What this reveals is the trading of our agency for the constant quick fixes of real-time 'knowledge'. It is a pathological obsession that legitimises the growing development of monitoring techniques and technologies. In this spirit, it reconfigures the very meaning of big data, which should now be defined at any point in time as 'data whose size forces us to look beyond the tried and true methods that are prevalent at that time'.[52] The scope of its use is further expanded into ever newer areas of social existence. Sports fans can thus draw on big data to increase their viewing choices, while coaches can exploit it for obtaining more sophisticated up-to-the-minute information about their players and opponents. Yet beneath these benefits lies the fostering of a profoundly unequal society between the 'data rich' and 'data poor'.[53] This mining of the self, moreover, is literally remapping the world, using algorithms to extract and disseminate knowledge about our shifting global geographies and the people inhabiting them. Just as with the cartographers of old, such data-based mappings are plagued by profound omissions reinforcing existing global power differences.[54]

Predictably, the ultimate goal of this data extraction is not benign. It is largely directed at discovering and maximising economic value for firms. Consequently, 'data silences or gaps result from the kinds of data deemed worth creating and storing. Simply put, corporate data is meant to create a profit, its veracity secondary to its economic value. In practice, this means that the everyday scale of data is the scale of the commodified data point and the individual person from whence it springs'.[55] The quickness in analysing such data, therefore, does not make up for what is already missing in these existent datasets. The veneer of 'objectivity' is undermined – though largely hidden – by the fact that the very collection of data is ideologically driven. Moreover, it decontextualises our lives and experiences, rendering them simply as data points to be studied, analysed and profitably exploited. The creation of so-called 'data doubles' epitomises this phenomenon, signifying processes of extraction by abstraction, whereby our identities are taken out of their 'territorial settings' and separated 'into a series of discrete flows' that can then be infinitely mined for information.[56] We are reduced to byte-sized datasets whose innovative monitoring and analysis reaffirm and reproduce existing inequalities and status quos. As such, in the present era it often seems that the more data we mine, the less critical knowledge we have about ourselves and the world.

Predicting Your Futures

If archiving is the making of your histories and data mining the real-time analysis of your present, then predictive algorithms and machine learning are the prophesising of your future. We are already witnessing the ability of big data to analyse what you have done and are doing in order to catalogue what will most probably happen to you.[57] The augurs of times past are currently the information signs of the twenty-first century. Superstition has been replaced by 'accurate' data forecasting. The buzz and activity of social media today presents a window into what may digitally come tomorrow.[58] We can immediately correct and refresh these predictions, using constantly updating data streams to reshape future expectations.[59] At a time when many fear our extinction, we remain more obsessed than ever about knowing in precise detail what fate has in store for us.

What this gestures towards is a profound contemporary human irony. The less control we seem to have over our collective destiny the more control we seek to have over our personal futures. Put differently, while as a society there appears to be little we can do to counter our inevitable decline, nonetheless we passionately turn to big data in order to recalibrate our own future possibilities. In turn, we see the dramatic change to our aspirations and even conceptions of morality brought about by the rise of digital technology and the emergence of virtual power. People are not so much consigned to a given outcome – either by society or providence. Instead, they are responsible for conceiving of exciting potentialities and continually monitoring themselves to ensure that they come to fruition. In the process, we are reprogrammed to become 'data time travellers', exploring fresh virtual realities and ensuring that they are profitable.

In this spirit, it is through such constant monitoring that we are able to make our lives at least partially our own. Virtual environments such as Second Life provide the opportunity for people to explore different lifestyles and personal possibilities. What if instead of being unemployed I was actually a doctor? Or what if instead of a lawyer I was a criminal? Implicit in these explorations is the idea that while our 'real' future may be rather limited and largely fixed in place, our virtual futures are wide open. Tellingly, these infinite possibilities create new opportunities for power holders to surveil subjects, using their fantasies of a different existence as prime fodder for better monitoring their hopes, dreams and preferences. Yet, they also catalyse the creation of 'counter-technologies' by users to protect their real and virtual selves from these secretive intrusions, safeguarding their present identities and future selves in the process.[60]

Moreover, while the predictive capabilities of big data may at first glance sound restrictive, they are actually portrayed as opening people and organisations to infinite possibilities. Business intelligence rests on the 'myths' of being 'data-driven, predictive and proactive, (having) shared accountability, and inquisitive'.[61] Data is transformed from an arbiter of what will definitely be to a resource for shaping what may potentially be. It is a compelling belief that through greater self-monitoring and algorithmic forecasting we can cast and recast our future in real time. The more we know about the present the clearer our fate becomes. The collection, cataloguing and constant refreshing of our life data is then a doorway to possible futures, an exploratory venture into the realm of potentialities. It also leaves the window even further open for peeping corporate and government eyes to surveil these virtual selves.

It is this freedom to use data monitoring to venture forth virtually into the future that masks just how influenced and ultimately overdetermined we are by these digital techniques. In addition, to secretly – and, of course, not so secretly – externally monitor our data they also prime us to accept the authority of data. Indeed, in a recent testing of a biosensing technology, individuals were given the opportunity to respond to clothing that displayed their emotions based on their collected personal data. Interestingly, people rarely if at all questioned the legitimacy and objectivity of these wearable data-monitoring technologies. Perhaps even more troubling, when given greater control over identifying their emotions, some people expressed insecurities over 'how much feeling they had'.[62] This uncertainty, this tension between wanting the agency to monitor our data but the security in having its results be determined for us, reveals pointedly the ways our lives, feelings, beliefs and actions are willingly dictated by data. In a sense, we embrace the opportunity to be data explorers in return for which we are told 'objectively' who we are and can possibly be.

Virtual Progress

Big data has made possible futures that are personalised, empowering, supposedly safe and utterly exploitative. It is a cross between classical tragedies like Oedipus and free market fantasies of working hard to make your dreams come true. Rather than asking if we can escape the fates, it is now more appropriate to query if we can escape our data-analysed destinies. While these are fixated on the individual, they also increasingly apply to collective aspirations. It is directly challenging ideas that we are at the dead-end of human history by revealing how data can reveal new possibilities for our shared improvement. In doing so it has encouraged fresh virtual potentials for the marketisation of our virtual futures and our common exploitation.

This casts new light on the notion of modern progress. While this idea may seem timeless, as the great historian J. B. Bury presciently reminds us, it is relatively new, really taking off in the sixteenth century.[63] It reached its zenith in the West in the nineteenth century and the elite-level optimism of the industrial revolution. However,

> While the twentieth century is far from barren of faith in progress, there is nevertheless good ground for supposing that when the identity of our century is eventually fixed by historians, not faith but abandonment of faith in the idea of progress will be one of the major attributes. The skepticism regarding Western progress that was once confined to a very small number of intellectuals in the nineteenth century has grown and spread to not merely the large majority of intellectuals in this final quarter of the century, but to many millions of other people in the West.[64]

These ideals of progress have, moreover, been critiqued for justifying present-day injustices and inequalities in the name of a fantasy of good times to come. Whether speaking of global development strategies, colonial attempts at 'civilisation', or even communism, the current struggles are always legitimised by a utopian vision of soon-to-arrive progress. It further orients humanity to think in linear terms of a past leading confidently into a better future, thus reinforcing the ideologies guiding this supposedly inevitable upward trajectory. Recent theorists have offered instead a post-structuralist conception of progress, one that promotes 'a more radically open-ended, futural conception of freedom, where we leave open the possibility where there may well be some future in which our own normative commitments and ways of thinking and ordering things have been transcended and thus will have come to seem impossibly strange'.[65]

Big data society has produced its own similar but sinister post-modern version of progress. It is one that portrays itself as 'radically open', proclaiming that virtually anything is possible, the only limit being our imagination. There is a hidden clause though – two actually. The first is that any future you pursue, regardless of how radical or challenging to the status quo, will ultimately all be raw digital resources for further data extraction and mining, and as such more information for the powerful to economically and politically profit from. The second is that this radical openness is closed to all possibilities that are not 'fiscally' sustainable. The imagined future, in this sense, is an opportunity for individuals and communities to come together and creatively conceive of ways to save capitalism, to make it adaptable to any and all of the new technologies, cultural values and radical rearranged social relation that may emerge.

This reveals, in turn, the digital construction of the progressive subject 2.0. They are tasked with using all available technological resources to make the world both better and more profitable. This is tied up with the use of self-tracking technology to continually monitor yourself and others to ensure that you are heading towards the future you desire.[66] Market rationality and digital sophistication combine to manufacture the economically virtuous individual, a literally 'informed' entrepreneur for creating a better tomorrow, starting today.

The quantified self is transformed through the technological magic of big data and hidden algorithms into the 'virtual self'. They must dutifully collect and intelligently turn their data into fresh opportunities for achieving personal fulfilment, professional success and social justice across their lifetimes. Data becomes an infinitely renewable resource for this very purpose. Here, 'data capital' is considered non-rivalrous, non-fungible (cannot be substituted) and experience-based, requiring greater data equality, data liquidity and data governance. Data discovery allows you to see previously unseen connections and perspectives, which can then be 'pushed aside' when new questions arise.[67] At stake is not merely improving your current circumstances but planning ahead to maximise the positive social and economic impact of all your possible futures.

Thus, while progress appears to be wide open it is in fact continually directed both individually and collectively to always reflect neoliberal values. The 'data revolution' is being driven by different sets of discourses mixing openness (e.g. 'sharing, reuse, open access', etc.) with market-based values (e.g. 'transparency, accountability, social entrepreneurship, and economies of scale'). Government and big business have adopted a smart 'rationale' focused on using big data in all domains of existence around the principles of 'governing people, managing organisations, leveraging value and producing capital, and creating better places'.[68] It is the personal responsibility of everyone to conceive and help create a more efficient, 'open', and marketable brave new digital world.

Monitoring Your Neoliberal Lives

This progressive form of monitoring legitimises a more expansive surveillance culture for guiding all of one's virtual prospects in line with neoliberal prerogatives of productivity, efficiency and profitability. In doing so, big data adds a new wrinkle to this regulation of capitalist time. It does not just seek to direct the course of a person's life but the lives of their multiple and diverse online selves. Firms and governments are constantly surveilling your different virtual personas in order to discover fresh ways to exploit them. As such, a novel 'datafication model' is beginning to emerge, where 'new personal information is deduced by employing predictive analytics on already-gathered data'.[69] In doing so, it mines and subtly – often even subliminally – suggests personalised lifespans that would be valuable to them.[70]

Mobile technologies, in particular, allow those in power to track your every move, and increasingly thought. Our texts, posts, phone conversations and online purchases leaves 'digital breadcrumbs' for prying digital eyes to learn more about us to enhance their control over us.[71] Uncovered is a novel type of 'reality mining'[72] in which

> Revolutionary new measurement tools provided by mobile telephones and other digital infrastructures are providing us with a God's eye view of ourselves. For the first time, we can precisely map the behavior of large numbers of people as they go about their daily lives ... [and they have] given us a new, powerful way to understand and manage human groups, corporations, and entire societies. As these new abilities become refined by the use of more sophisticated statistical models and sensor capabilities, we could well see the creation of a quantitative, predictive science of human organizations and human society. At the same time, these new tools have the potential to make George Orwell's vision of an all-controlling state into a reality. What we do with this new power may turn out to be either our salvation or our destruction.[73]

Crucially, this hi-tech Big Brother is not just watching what we are doing now, but predicting and seeking to dominate what we will go on to do.

These predictions form quite daunting and inscriptive digital expectations. They point to how, with only a little more effort and discipline, we could achieve our goals and become our best selves. It is a simultaneous means for asserting both 'the control of the self and self-control' whereby '"algorithmic governmentality" relies on the dream that reality, if correctly probed and recorded, will reveal its own passive, inoffensive and non-coercive normativity ... The new social outcast could very well become the one who is unable or unwilling to be "oneself", and who by denying the "objectivity" or "undeniability" of one's traces, is denying one's "self".'[74] Across our diverse 'life-plans' we must constantly endeavour to live up to what our data overlords say we can and should be.

And of course, this digital control is predictably oriented towards a market-friendly vision of your and society's future. They voluntarily and involuntarily guide our personal and social development.[75] All your actions, behaviours, random shared thoughts and online preferences become potentially weaponised as a digitally means to judge you against these exacting and never-satisfied neoliberal standards. Our 'digital footprints and data fumes' act to inscribe 'certain meaning into quantified spatial information'.[76] These become means by which to retrospectively and prospectively hold people accountable as market subjects.

Under these conditions, new selves compatible with the needs of twenty-first-century capitalism are painstakingly predicted and produced. Beneath the veneer of an exciting data-driven future are much more insidious and controlling digital realities. Our lives and dreams represent constantly updating opportunities for disciplining and exploiting us. Our existence is increasingly 'motile', the mobile and flexible shaping of our short- and long-term virtual possibilities. Hence,

> The 'you' being sold – the social data we all generate – is motile; that is, it flows from us, through our myriad personal technological artefacts and the material intricacies of the cloud initially as an expression of sociality. Yet its movement is not directed by us, and is almost wholly autonomous of our control. Indeed, the data we generate increasingly is moving at the behest of capital and the state.[77]

Underneath these consumable present and future identities is a sophisticated and dangerous real-time surveillance regime. Perhaps even more troubling, it uses this omniprescient data-mining technology to end certain lives while saving others. Consequently, 'The technologies of analytics focus human attention and decision on particular persons and things of interest, whilst annulling or discarding much of the material context from which they are extracted'.[78] By turning us into easily digestible and exploitable data bytes we can be moulded now and forever, into whatever virtual reality we 'choose' to inhabit, into a prime source for digital control and exploitation.

Smart Times

The advent of the big data era has created new aspirations for how we envision and lead our lives. We increasingly no longer cast our fate to the proverbial winds or simply hope that hard work will pay off. Instead we assiduously collect, catalogue, analyse and act upon our data. We embrace the prediction that hidden algorithms working behind the scenes are objective and true. We track our fate daily and change our behaviour accordingly. This uncovers a new fantasy of data management that spans our various online and offline lifetimes, where we are in control of our information and can constantly exploit it to our advantage.[79]

This is reflected in the evolution of human history in line with this 'datafied' post-modern reality. Specifically, it revolves around the desire for future psychic wholeness associated with this fantasy of data management and monitoring. People continually strive to realise the visions of their perfect self held out to them by smart analytics. They work diligently and regularly to match their current data selves to the projected virtual identity they so desperately would like to be one day. In a disaggregated technological world, this digital pursuit provides ontological security and a stable identity – one inexorably linked to the affective promise of big data. This echoes Lacan's description of history as the circular and ultimately repetitive striving for an elusive perfection and harmony.

At the heart of this history is a fantasy of infinite information, limitless personal development and unending social progress. It is the insatiable need to learn ever more about ourselves and our community with the hope of constantly improving them. As such, 'Big data is about more than just communication: the idea is that we can learn from a large body of information things that we could not comprehend when we used only smaller amounts'.[80] The bigger obviously the better, in this regard. This plays on the mythical quality of data analytics for our lives. Indeed, algorithms are so much more than mathematical functions or technical solutions. They are almost magic-like entities whose sorcery can reach deep into our present and foretell our future. Their 'social power' is in their ability to dramatically alter our 'broader rationalities and ways of seeing the world'.[81] They make us feel as if we are living in 'smart times', where all the information we need to make our dreams a reality is in the world and inside of us.

Yet scratching beneath this dazzling promise of big data living is a more anxiety-filled and disciplining existence. Big data is not just our saviour but our hi-tech prison guard, tracking our every move and judging everything we do. As leading technology scholar Kate Crawford presciently observes, 'Already, the lived reality of Big Data is suffused with a kind of surveillant anxiety – the fear that all the data we are shedding every day is too revealing of our intimate selves but may also misrepresent us'.[82] Moreover, this unease is not confined to those being surveilled, it also applies to those responsible for using big data increasingly to watch our every move. It is what Crawford refers to as the 'anxiety of the surveillers', denoting how 'no matter how much data they have, it is always incomplete, and the sheer volume can overwhelm the critical signals in a fog of correlations'.[83]

These reveal the presence of a datafied 'big other', who does have complete data knowledge and who therefore has completely monitored themselves to their own benefit. Just as visceral and pressing, though, is the feeling of being overwhelmed, of wanting big data to make sense of our information-saturated society and tell us what we should do. It is about using big data technologies and methods in order to simplify our environments and give coherence to our diverse online existences. This feeds into deeper 'anxieties of control', signifying the 'desire to discern (be aware of) and direct (determine the disclosure of) personal spatial big data flows about oneself while feeling that any attempt at exerting such control is effectively futile'.[84] In this sense, big data is both the perpetrator and solution to our feelings of being out of control.

This elusive fantasy of exerting digital control over our lives has the potential to reframe contemporary understandings of inequality. Rather than focus on conventional material and social disparities in resources and power, it now concentrates on the differences people have in drawing on big data to 'intelligently' improve their lives and decision making. It represents a 'new kind of digital divide: a divide in data-based knowledge to inform intelligent decision-making'.[85] This longing to be ever 'smarter' in our choices drives us perpetually forward as 'data subjects'. It propels us into tracking, analysing and exploiting more of our data as the way to become the ultimate controller of our digital fates. It assumes that there is a right way, a perfect choice, an optimised life that can be discovered if we only had more information.

This desire allows us to ignore the concrete and virtual ways big data rapidly and often invisibly takes over our lives. We gain a distinct enjoyment from remaining cynical about either the 'excesses' of dataveillance or its futility – thus ignoring the totalising effect of big data for creating the 'datafied' subject. In this respect, it appears too large or too ineffective to properly monitor or control.[86] This cynicism and parallel embracing of its empowering possibilities, permits elites and the capitalist system itself to remain unaccounted for and unaccountable even while they are increasingly shaping and disciplining our existence. People are well aware that they are being watched, that their data is being used against them, that algorithms are directing their choices. Yet they are met progressively with an apathetic shrug, a feeling that this is the price we must pay for achieving better living through data – that this is the acceptable dark side of data progress and getting to exist in smart times.

Planning Your Life at the End of History

Free market democracy was meant to be the 'end of history' – the final epoch of human achievement after millennia of wars, domination and exploitation. The spectre of the financial crises and the potentials offered by big data have upended this once triumphant historical conclusion. Yet the future seems as dark as it does bright. Dystopian visions of continued corporate rule, technological dehumanisation and even worse inequality, threaten rosier promises of progress brought to us by 'smart' technology. Compounding these fears on the horizon are present-day concerns that we are being invaded and controlled by big data. Amid these existential worries is perhaps a more profound change happening in real time. Digital technology is rapidly transforming how we conceive of and live our lives. It permits us to archive our past and monitor our present for realising a better personal tomorrow. Our data-measured actualities hence provide the foundation for our exciting virtual futures.

This culture of monitoring associated with emerging modes of 'virtual power' has made planning and accounting for your own personal future central to the empowerment and control of contemporary subjects. Subjects utilise monitoring techniques and technologies to manage their personal 'life histories', charting and predicting how to make their short-, medium- and long-term actions as efficient, productive and profitable as possible. This eternal accounting for one's existence fosters a sense of hope and agency within a perceived general 'end of history' in which there can be no society beyond that of the free market. In doing so they open not only new hopeful possibilities for themselves but also for their disciplining and exploitation spanning their real and imagined lifetimes.

7
Totalitarianism 4.0

How good a person are you? Do you make a positive or negative impact on your community and society? These are questions that most people often ask themselves. The answers have been, until now, largely subjective and unmeasurable. However, China is aiming to completely reboot the concept and practice of private ethics and the public good. In a move some have ominously referred to as 'big data meets Big Brother', its leaders are seeking to implement a revolutionary 'social credit' system for all of its citizens:

> Imagine a world where many of your daily activities were constantly monitored and evaluated: what you buy at the shops and online; where you are at any given time; who your friends are and how you interact with them; how many hours you spend watching content or playing video games; and what bills and taxes you pay (or not) ... now imagine a system where all these behaviours are rated as either positive or negative and distilled into a single number, according to rules set by the government.[1]

While foreign critics fear for the worst, the Chinese government portrays this as an innovative part of a modern 'socialist' society. This data-based 'credit' system, they proclaim, is imperative to 'establishing the idea of a sincerity culture, and carrying forward sincerity and traditional virtues' by incentivising people 'to keep trust and constraints against breaking trust as incentive mechanisms, and its objective is raising the honest mentality and credit levels of the entire society'.[2] In practice, a 'bad social credit score' can lead to being banned from travel, slower internet speeds, your kids being rejected from the best schools, being prevented from getting good jobs, being barred from the best hotels and public shaming as a 'bad citizen'.[3] Some citizens in cities where it has already been introduced have already seen its positive effects. As 32-year-old entrepreneur, Chen, proclaimed: 'I feel like in the past six months, people's behaviour has gotten better and better. For example, when we drive, now we always stop in front of crosswalks. If you don't stop, you will lose your points. At first, we just worried about losing points, but now we got used to it.'[4]

Despite these positive reviews, this system raises profoundly worrying questions about the future of human freedom. At the most basic level it 'Puts its people under pressure to be model citizens'.[5] It represents a potentially less chaotic but just as inscriptive twenty-first century digital cultural revolution. And it is one that is not just confined to China. As the famed technology theorist Adam Greenfield notes, 'there's nothing so distinctly Chinese about it that it couldn't be rolled out anywhere else [should] the right conditions obtain. The advent of social credit portends changes both dramatic and consequential for life in cities everywhere – including the one you might call home.'[6]

However, in principle it could also serve as a radical opportunity to use big data to make businesses and governments more accountable, as their data are also being publicly tracked and shared. In practice, though, there lurks a much more exploitive reality. It is not a coincidence that these policies are being tested in poor areas such as Guizhou, whose low regulations and general lack of public visibility have attracted tech giants including Google, Microsoft, Bidu, Huawai and Alibaba, all of whom have 'established research facilities and data centres in the region', with Apple soon to follow later in 2018. In this respect, 'Guizhou's position as the country's data centre makes it an ideal social laboratory for the local government's Social Credit System experiments'.[7] What this has uncovered is the attempt by the tech industry to combine its insatiable desire for more data to exploit with a compelling, if rather authoritarian sounding, picture of social advancement and cultural goodness.

The Chinese credit system is just the latest and most obvious threat of a fully operational and marketised form of digital authoritarianism. It is a contemporary repression built on the fact that increasingly all aspects of our life are now transformed into quantifiable data[8] and are therefore prime resources for our monitoring and control. For this reason, 'Many countries are in the throes of a debate about the amount of surveillance a government should be allowed to carry out on its own people. But in other countries, where there are few, if any, checks on the state's powers, a potential dictatorship of data is already on the horizon.'[9] Equally troubling are the ways in which 'big data, big brother, (and) big money'[10] are connected with increasing intricacy and driving this 'smart' repression.

These policies, and what may realistically follow them, once more raise profound questions concerning the role of algorithms for human governance. They serve as a type of 'myth' that secretly controls our lives.[11] They have a growing influence on the regulation of our relationships and societies on a previously unthinkable scale.[12] Instead of deploying data to survey the world they are now directed more and more at 'mapping' the entirety of our lives, actions and beliefs.[13] And it will do so not from a prison watchtower, a government installed TV screen in your home or even a surveillance camera on the street, but primarily from the palm of your hand using an everyday technology that you have voluntarily installed and more often than not bought and paid for.

The new Big Brother is not the stern-looking dictator looking down at you from films and posters but an omniscient digital 'cloud' that operates by 'continuously, securely, and privately analysing the digital traces you generate as you work, shop, sleep, eat, exercise, and communicate'.[14] This invasive surveillance is aided by the constant 'digital tracks'[15] we are leaving as we carry on with our online lives and being economically exploited by capitalist firms as well as their government allies.[16] Yet not all hope is necessarily lost. These intrusions into our freedom and privacy are breeding their own forms of resistance. Countering such intrusive surveillance are types of 'sousveillance' in which people repurpose this tracking and watching technology in order to better 'watch the watchers'.[17] There are now products like 'haccessible' glasses that record your movements so that you can establish 'digital alibis' in case you are falsely accused of a crime.[18] More politically, mobile phones have allowed everyday citizens to film police and politicians in order to reveal to the wider cyber-world their abuses of power.

Still, this updated tale of authoritarianism and its committed opposition tells only part of the story. There is an even darker underbelly to these encroaching data oppressions: how algorithms will enact authoritarian models of decision making in our everyday life outside of human control. In fact, the function of human power will be to coercively enforce these 'objective' decisions from our data rulers and police those who would dare resist. If this sounds far-fetched then consider that this is already happening at a small scale, where, for instance, airlines use sophisticated algorithms to determine when to throw people with tickets off planes due to overcrowding – commonly in ways that defy 'common sense' but which still cannot be challenged by either employees or customers. Importantly, 'Once initiated, authoritarian decision making (comply!) can result in violent escalation'.[19]

It also feeds off an ironic but genuine desire by contemporary citizens to meet the imposing challenges of a world where global capitalism appears to be creating more problems than benefits for the majority of the people in the world. Hence, 'Facebook, Google and the "big data" revolution is undermining Western democracy while strengthening the hand of authoritarian states to address the challenges thrown up by globalisation'.[20] In particular, it revives the idea that the market can be planned, positing big data as the key to properly coordinating and efficiently organising economic relations. It is proclaimed that 'the explosion of data in our modern world could – at least in theory – inform far better managerial decisions and reduce the information imbalances between a planned and a market economy. Central planners are rapidly acquiring the tools to process data a lot more effectively.'[21]

This chapter outlines in stark detail how digital technologies pose a new political risk of totalitarianism. Notably, it paints a picture of a troubling near future where everything we do will be fully monitored and judged as to its economic value. It will also reveal how this 'totalitarian 4.0' will allow elites to exploit new future tech to further evade personal accountability while promoting the continued systematic irresponsibility of capitalism. It proposes that we are moving into an era beyond 'surveillance' or 'sousveillance' towards 'totalveillance', where we are all seeking to watch each other all the time and compete with one another over who can profit the most from this emerging total surveillance culture. It will conclude by analysing the all-encompassing and total character of 'virtual power' – one that is increasingly depersonalised and extends infinitely into all our virtual and physical interactions.

Closed Intelligence

The twenty-first century was meant to be the golden age for democracies. The fall of the Soviet Union was viewed as the beginning of the end for autocracy and dictatorship. While clearly this could not happen overnight, it was expected that free market liberal democracies were the inevitable wave of the future and all past forms of tyranny would be washed away in its high tide. These dreams have turned into a living nightmare in only a few decades. The first part of the new millennium has been wracked by authoritarianism and illiberalism. Much of this is a reflection of the fact that capitalism, despite optimistic claims that it is the only sure road to democracy, lends itself to state repression and cultural violence as well as oligarchy and profitable mass incarceration.[22] However, digital advances have added a new twist to this virulent reliance on technocracy and oppression, positioning data as the ultimate source for 'smart' governance with little need for human intervention. Its 'objective' knowledge relies on techniques of 'predictive analytics' and 'real-time policy instruments' to reshape authority at both the macro and micro levels.[23]

A claimed benefit of this digital rule is that it would allow for dramatically increased transparency. The prospect of 'open data' has obvious appeal in a society that can seem to be relatively closed and where information is power. It would supposedly set free our information from behind its closed digital walls, thus giving us an unvarnished and better insight into ourselves and society. Yet such disclosures are not always as 'open' as we might like to believe. Critically, 'mediating technologies, conceptualized here as disclosure devices, have distinctive organizing properties that are important to scrutinize. They play a central role in attempts to shed light on objects, subjects and practices, and to help build or break up relationships within and across sites and organizations'.[24] Thus, how and why data are made 'open' is crucial for shaping contemporary social relations and discourses of transparency as well as ultimately accountability. Moreover, something else is at stake: who are the 'data intermediaries' helping to bring about this openness and what are their motivations and ideologies?[25] It is extremely telling that while governments preach the value of 'openness' they have prosecuted with a passion data whistleblowers such as Edward Snowden, who have revealed the depth and scope of our current state-backed surveillance society. These concerns are compounded by the looming threat of a 'digital dark age' in which huge swathes of online information could become lost and corrupted due to being stored on outdated systems.[26] In the present, therefore, it is always imperative to ask what data are being brought to light and what are being kept in the digital shadows.

Much more important than gaps in our digital knowledge is how data dresses up human bias and inequality in the hi-tech clothing of scientific 'objectivity'. Even at the most basic level, biased sampling often skews big data due to widespread issues of self-selection.[27] They ignore, in this regard, the human element – who is collecting their data and how is this sharing tainted by their own explicit and tacit prejudices? Going beyond mere biased samples, algorithms reproduce in their learning and analysis systematic assumptions of racism, sexism, homophobia and classism.[28] The intelligence it thus provides perpetuates these systematic biases, and even strengthens them through supposedly being ideologically neutral and free from human interference.

Predictably, big data has been politically and socially weaponised as part of the broader control strategies that target historically marginalised populations. It feeds into, for instance, the range of barriers poor people face in accessing social benefits and bettering their economic condition.[29] In the present era, it is fair to say that data contributes to the rich getting richer and the poor, tragically, getting poorer. It also is directed at surveilling minority groups in order to discipline and regulate their behaviour. This continues a longer history of employing technology for maintaining the status quo and power differences. As Malkia A. Cyril, the founder and executive director of the Center for Media Justice, presciently observes:

> Early technologies, and the policies and practices that undergird them, were forged to separate the citizen from the slave. The slave passes, branding, and lantern laws of then have become the cellphone trackers, facial recognition software, and body-worn police cameras of now. Their mission, however, hasn't changed much – to catch and

> control black dissidence – only now they're doing so in a digital age. These technologies have been incorporated into the law enforcement process at every level, from predictive algorithms for assessing pre-trial risk and criminal activity to widely adopted police technologies that face little to no oversight. These technologies – including cell-site simulators and surveillance cameras – are trained on communities of color, especially blacks, immigrants, Arabs, and Muslims. In each case, the presence of technology, of math, is touted as the lynchpin for countering bias despite clear evidence that data derived from discriminatory processes reinforces, not eliminates, bias.[30]

What this reveals is the rise of a progressively and intentionally oppressive form of data intelligence. The goal is to capture as much data as possible and to use the information as strategically as possible for purposes of domination and profit. This reflects a refreshed arms race revolving around who has the technology to collect and effectively use the most data, and in turn represents a new type of surveillance. It is one that is 'structurally asymmetrical' since it 'is available only to those with access to the databases and the processing power ... likewise the forms of knowledge it generates are necessarily opaque; they are not shareable understandings of the world, but actionable intelligence crafted to serve the imperatives and answer the questions of those who control the databases'.[31] Consequently, we have evolved from the looming apocalyptic threat of 'weapons of mass destruction' to the hidden but no less dangerous 'weapons of math destruction'.[32]

Real-Time Tyranny

Big data is not only updating traditional authoritarianism but rapidly speeding it up. Indeed, politics and domination are to a large extent a matter of timing. Different intellectual and ideological epochs, for instance, can be accurately assessed as 'temporalisations' in their shaping of how people understood and experienced time.[33] Concretely, giving social and governing processes deadlines and schedules can have dramatic 'political value' for a status quo.[34] The present age is progressively trapped in 'virtual time'. It is flexible and customisable to each individual's needs. Yet it involves constant real-time surveillance and monitoring to be sure that it is always being used productively and profitably.

Governance itself is quickly leaving human time behind. Traditionally, sovereign power was built on rulers having clear geographic and temporal boundaries. Referred to by Foucault as 'étatisation', it was the creation of a definite state led by a ruler who was tasked with making 'strategic' and 'cunning' short- and long-term decisions on its behalf. The advent of modern democracy further institutionalised time around regularised elections and bureaucratic governing processes. The current demand for constantly updating data and the 24/7 pressures of an 'always on' global capitalism have made such timings seem antiquated. Instead, instantaneous and predictive mechanisms based on 'objective' information are required, for making important decisions both quickly and wisely.[35] The digital era therefore needs a post-human form of rule that is largely independent of human oversight. To this effect, 'Big Data approaches have the potential for developing self-governing societal capacities for resilience and adaptation through the real-time reflexive awareness and management of risks and problems as they arise'.[36]

The near future of governance will be a refreshing technocracy. Refreshing in the dual sense of being regularly updated based on real-time data analysis and partially quenching our insatiable thirst for ever more information. 'Smart' governance depends on a desire for real-time data that leads to 'technocratic governance and city development, corporatisation of city governance and technological lock-ins, buggy, brittle and hackable cities, and the panoptic city'.[37] This rule by algorithm can surveil our individual actions in order to eternally monitor and direct our collective development. The human biases driving this 'intelligent' governance, in turn, 'objectively' reproduce existing inequalities and systematic forms of discrimination. Democracy, in turn, is thrown by the wayside in the name of 'technocratic values and uneven development'.[38]

However, this data-driven autocracy does much more than oversee public policy. It also holds the potential for guiding our conduct as digital citizens. It represents the prospect of automated disciplining where individuals and communities are guided through a range of hi-tech and often hidden cues to embody certain desired values and actions. This subtle but forceful governing of human behaviour is, of course, not new in itself – reflected in the past, for instance, in the modelling of the school schedule on the capitalist workday, including a similar bell telling students when each activity is done just as if they were factory employees.[39] Yet the contemporary period differs in one key respect. It is trying to imperceptibly use citizens to experiment with different sets of social existence and timings to maximise efficiency and profits. What is being imagined is a type of 'Walden 3.0' – when 'privacy issues are framed as "control over information" it becomes apparent that some areas in the digital world might be heading to what I call Walden 3.0; communities of interest that are influenced and controlled by measurement and experimentation'.[40] The deployment of big data to help resolve traffic congestion or improve health care waiting lines is certainly potentially welcome. However, they also invite our exploitation as test subjects for gathering ever more information for our exploitation.

Significantly, this automated politics plays on growing popular desires for a more 'open' society and greater social visibility.[41] While the algorithms may be hidden, the data it relies upon are potentially available for all to see. It thus provides the veneer of transparency, replacing public deliberation and democratic accountability with a legitimacy based on digital visibility. It is accepted that data may run our countries and rule our lives. Yet the information being used to do so must be open and clear. This same ethos, in turn, is justifying public forms of digital shaming and discipline. Returning to the example of the Chinese 'social credit system', those whose data show they are guilty of 'bad citizenship' will be 'openly' condemned via social media and billboards. While this may seem extreme, to a certain extent it is already occurring economically across the world, as 'bad credit scores' can be used to bar people from getting a mortgage, buying a car or even taking out an additional loan to further their education.

What these 'open' types of disciplining represent, in turn, is the shifting of public responsibility from the system to the individual. A bad credit score is not the fault of a rigged economy that forces us to take unfair loans simply to survive. Instead they are a reflection on our own personal inadequacies – our inability to properly manage our finances. Data is deployed, in this respect, as a tool for 'scientifically' proving whether one is a good market subject and therefore deserving of punishment or reward. Similarly, the socially manufactured desire for 'post-human' governance masks the real human interests behind this hi-tech technocracy. It allows elites to hide their manipulation of the economy and politics to their advantage as 'intelligent decision making'. Additionally, it enhances their ability to flexibly update and adapt their power to changing social circumstances, constantly surveilling social trends to prevent any 'viruses' that would threaten their privilege and dominance. While these efforts will always be necessarily incomplete, they nonetheless represent the attempt to establish a form of real-time tyranny.

Totalveillance

The dream of every totalitarian government, and nightmare of almost all its citizens, is to be able to completely surveil everyone and everything within their official boundaries and beyond. In a sense, this is simply taking sovereign modes of power to its most extreme form. Notably, the figure of the Machiavellian prince is one who has full knowledge of his subjects so that he can protect them and maintain power over them. The driving force of totalitarianism may be raw power, but its underlying justification is having the information required to rule efficiently and wisely. However, big data has made this totalising desire closer to a reality than perhaps ever before. It encourages a monitoring culture where people personally track themselves and voluntarily surrender it to elites.

To this end, surveillance is not only dramatically expanding but being utterly transformed. The conventional top-down mode of watching and regulating people is being exchanged for one that is much more omnipresent, distributed and shared. Authority and control now depend on 360-degree surveillance. Individuals increasingly monitor themselves and each other, which is made possible through digital technology and big data. This directly relates to the valorisation of 'productive time' within contemporary capitalism – as each moment becomes a new opportunity to get something else done quicker.[42] We are now engaged in digital prosumption, where we produce and consume social media content and online purchasing simultaneously, monitoring the experiences of others and shaping our own behaviour based on this shared information.[43] Consequently, we have become boundaryless surveilled subjects in which we are 'simultaneously always a producer, always a family member and always a consumer',[44] being equally tracked and monitored in all these roles.

This reveals the potential 'totalisation' of surveillance. Put differently, it is now theoretically conceivable that the entirety of our private and shared lives can be collected, catalogued and analysed. For perhaps the first time humans are truly confronted with the threat 'of the end of privacy'.[45] While this is certainly still a long way off, and a powerful myth in its own right, it still reflects a dramatic shift in the relation of surveillance and power. Notably, processes of totalisation usually has the goal of instilling ideological conformity. The totalising aspect, in this regard, is the need for people to accept, internalise and follow dominant values. However, in the digital age the aim is much more subtle and disciplining. Instead it is to encourage virtually any and all activities in order to collect more data and find new ways to exploit and profit from individuals. Power, in this respect, is now more than simply productive – it is outright creative. The prospect of 'total surveillance' encompasses both what we do and everything we can possibly imagine doing.

Yet if these totalistic monitoring regimes encourage exploration, so to does it breed creative forms of digital resistance. Social movements, in particular, have engaged in covert actions to reduce the ability of those in power to monitor their behaviour and track their movements:

> [A] mutual relationship between resistance and surveillance unfolds as one side reacts to the practices of the other: as soon as activists advance in the protection of their contents of telecommunication, the surveilling parties concentrate on meta-data to explore the whereabouts of their targets. To counter this threat only the discontinuation of mobile phone use has been articulated.[46]

Still, such resistance is always paradoxical in its effect. The greater individuals innovatively struggle against such monitoring, the more they reveal to those in power new ways to overcome such data subversion and fresh opportunities to enact digital control over the population at large.

Significantly, the expansion of surveillance is pushed forward by its increasing depoliticisation. Whereas the conventional account of totalitarianism is one of a hyper-politicised society full of persecution campaigns and the need to be ever vigilant against state 'enemies', in the digital age it is premised more on simply helping people and societies to manage their affairs more efficiently. Its ostensible goal is not to 'take over the world' or 'remake society in its image'. Instead, 'operations of collection, processing and structuration of data for purposes of data mining and profiling' are created with the intention of 'helping individuals and organisations to cope with circumstances of uncertainty or relieving them from the burden of interpreting events and taking decisions in routine, trivial situations [that] have become crucial to public and private sector activities in domains as various as crime prevention, health management, marketing or even entertainment'.[47] It acts to make sense of a disaggregated and confusing contemporary reality while also making it easier and safer to live in. It masks its political and exploitative intent under the guise of a friendly global neighbourhood technology that simply wants to be 'helpful'.

Consequently, it adds an extra layer of authoritarianism to neoliberalism. Here personal responsibility is partially exchanged for the expert rule of algorithms. This hi-tech technocracy is invading all aspects of present society. Tech companies, for instance, are seeking to take over public education:

> These new schools are being designed as scalable technical platforms; funded by commercial 'venture philanthropy' sources; and staffed and managed by executives and engineers from some of Silicon Valley's most successful startups and web companies. Together, they constitute a powerful shared 'algorithmic imaginary' that seeks to 'disrupt' public schooling through the technocratic expertise of Silicon Valley venture philanthropists.[48]

It is similarly infesting governments and firms, strategically portraying data as an 'objective' decision-making tool for pushing forward total digital control.[49]

We are entering the age of 'totalveillance'. It is total in several important and profound ways. It seeks to monitor the totality of our experiences – both actual and virtual, real and imagined; it claims to have a total monopoly on objectivity and truth; and it is seeking to totally conquer and reshape society as a whole. The only things not being watched, it seems, are elites and the capitalist system itself.

Good Data Citizens

The prospect of total monitoring is in part an end in itself. The ability to collect and analyse even more data reflects the insatiable character of capitalism rebooted for a digital new millennium. While natural resources are limited, data is supposedly infinite – making data a near perfect partner for capitalism's unquenchable desire for markets. Yet data also has a significant disciplining effect. The emergence of totalveillance – both in fact and idea – has produced market-friendly 'good data subjects', willing and able to be digitally controlled and exploited.

The introduction of data analytics as a prominent feature of surveillance has dramatically enlarged its disciplinary possibilities. In particular, it is able to predict how individuals will act and react to a wide range of environmental factors and pressures. As such, it can simultaneously deepen and widen its exploitative potential. It can better regulate and shape them as consumers and employees based on their current preferences and practices as well as forecast and shape how they might respond to changing conditions.[50] This predictive monitoring extends to governments and citizens. By being able to accurately foretell the movements and views of the population they rule, states can prepare them for 'shocks' and actively mould them to meet the needs and desires of a status quo.

This updated Big Brother is, hence, more proactive than reactive. It encourages individuals to use data for enhancing their own well-being. In doing so totalveillance becomes a precondition for their personal wellness and professional success.

There is a moral duty to others and themselves to use data and self-tracking technologies to 'do more with less' and find innovative, efficient ways to improve their positive impact in their communities and workplace.[51] Additionally, it sets the basis for 'what counts' within organisations – linking the supposed 'objectivity' of data collection and analytics to reified market values of productivity and profitability.[52] At the macro level, these same digital logics are employed to delegitimise and spread free market policies globally, nationally and locally under the banner of 'smart economics'.[53] Close to home, so to speak, are 'biometric body projects' meant to productively monitor our bodies, digitally disciplining them to maximise economic value and social health.[54]

While this monitoring is largely voluntarily, it does require enhanced and innovative forms of policing. One of the most interesting examples of these efforts is the current use of social 'bots' to influence online behaviour.[55] These cyber-'friends' encourage people to display good internet behaviour – from fighting online racism[56] to promoting greater civility on social media. Although this 'friendly' undercover internet policing of personal cyber-conduct seems innocuous, these bots can also potentially be used to reprogram people's preferences.[57] It is a hidden re-education of individuals to become more active and confirming good neoliberal 'data' citizens. This pervasive form of data policing is further legitimised, as required, for monitoring 'cyber-security' and as such keeping societies safe from digital attacks.[58]

Following all of us everywhere is the roving omnipotent 'digital eye'. It has supposedly rendered surveillance 'perfect' through creating an 'electronic panopticon'.[59] It tracks, predicts and guides our movements, daily activities and stated preferences. This new data-driven social order is an outgrowth of a broader 'inner panopticon' for disciplining 'workers who are removed from the immediate sphere of influence of management and co-workers'.[60] While to a certain extent real, this digital eye is also a dangerous fearful projection. Significantly, it is this popular terror of 'always being watched' that, just as with past totalitarianism regimes, makes up for its actual surveillance gaps and limitations. Totalveillance exists as a full-scale public syndrome, a collective worry that we are being socially monitored and judged.[61] This covers over the fact that governments and firms still have limited capacity to collect and analyse all these data. Yet the perception is often in itself to discipline people's behaviour at work and within society. Indeed, even the most privileged workplaces are turning into 'digital sweatshops' marked by digital forms of control.[62]

Yet such totalveillance is also accomplished through various acts of self-policing. This can come in the form of the use of physical and online 'life coaches' who rely on data and 'computer monitoring' to track people's progress and motivate them to maximise their personal goals.[63] Less explicit but every bit as impactful is the function of social media for regularly, often imperceptibly, regulating our conduct. Constantly logging online we open ourselves in real life to obvious and hidden forms of cyber-tracking and disciplining.[64] These digital control techniques are enhanced by our own self-monitoring to fit into and be accepted by diverse online networks. Tellingly, the sharing of our successes and failures plays into the creation of a common community of support meant to motivate us to further maximise our personal and professional success.

Perhaps, though, the most insidious aspect of this data policing is just how convenient it is becoming. Indeed, it is hard work safeguarding our online privacy and even harder actively subverting the gaze of our digital Big Brother. It is understandably tempting to just accept that various elites have our data and will largely be able to use them as they please. Not surprisingly, authorities themselves are wilfully exploiting these tendencies to their advantage. In addition to making it harder and more annoying to circumvent this seemingly omnipresent digital control, they are also making it much easier to voluntarily surrender to it. The US government, for instance, has created FAST (Future Attributes Screening Technology) – a programme developed by Homeland Security that seeks to use a person's attributes to test if they are a future terrorist. In particular, 'FAST uses non-contact sensors to remotely analyse physiological and behavioral cues including, eye movement, body movements and other factors that an individual typically does not consciously control. The system conducts real-time analysis of the data collected in order to develop an objective recommendation for secondary screening.'[65] In the present era we are not only being controlled by a secret data police, but increasingly we are knowingly paying for their monitoring services.

Enterprising Monitoring

There is little doubt that the digital Big Brother is keeping an ever closer eye on us. It surveils our online and 'in real life' selves. Its scope is ever expanding, invading both our existing and imagined realities. It reproduces widening material inequalities by increasing its virtual power over us. Both in actuality and in our popular imaginations we are becoming completely monitored.

Interestingly, the digital age was meant to usher in a more inclusive, open and participatory culture. No longer would the media be dominated by the traditional cast of characters. Social media would be democratising – giving fresh opportunities for regular people and traditionally marginalised voices to be heard. In practice, while there is certainly more diversity, there remains a shared underlying, often unintentional, ethos connecting this seemingly fragmented cyber-public – namely profitability. To this effect,

> The use of big data to inform media production causes problems in the public sphere not because it fragments public debate, but because it somewhat paradoxically recentres

> public engagement around the complementary interests of the broad majority and profitability. The problem for public engagement is not that there are no overarching or all-encompassing media structures anymore but rather that these systems are informed by algorithms that promote a particularly populist 'profitable and normal' media experience.[66]

In a capitalist refresh of Mao's desire to let 'a hundred flowers bloom', neoliberalism is happy to let the digital garden become overgrown so long as all of its various plants bear economic fruit.

The ultimate trick of achieving totalveillance is to remain flexible and opportunistic. It is not so much an attempt to build a monolithic surveillance regime, but rather an agile and constantly adaptable set of techniques for monitoring human actual and virtual activity. Hence, while 'Our era is one of increasingly pervasive digital technologies, which penetrate deeply into the very core of the products, services, and operations of many organizations and radically change the nature of product and service innovations', its 'fundamental properties' are ones of 'reprogrammability and data homogenization. Together, they provide an environment of open and flexible affordances that are used in creating innovations characterised by convergence and generativity.'[67] Conformity, in this respect, is found in similarly embracing the fact that we are being monitored in a diverse and evolving set of ways.

This entails reconsidering monitoring as if it were akin to a market opportunity, asking constantly 'what gaps exist in the monitoring market?' Here the logic of capitalism and social control merge into a neoliberal imperative to innovatively and profitably fill these surveillance gaps. Marx famously argues that within a market economy what were once merely preferences become social needs. The contemporary era reveals this, of course, in stark detail as computers and mobiles which were in the recent past considered personal luxuries have now become essential to leading a connected and successful contemporary life. Analogously, surveillance is shifting from a form of control, to a hidden part of a hi-tech luxury lifestyle, to a definite need that people have to further empower themselves through data. As such, surveillance has become a shared and dynamic 'prosumptive' activity between producers and consumers:

> The co-evolutionary perspective on algorithms as institutions, ideologies, intermediaries, and actors highlights differences that are to be found, first, in the growing personalization of constructed realities and, second, in the constellation of involved actors. Altogether, compared to reality construction by traditional mass media, algorithmic reality construction tends to increase individualization, commercialization, inequalities, and deterritorialization and to decrease transparency, controllability, and predictability.[68]

Consequently, the new byword of surveillance must be innovation. If the need for raw data is insatiable so is the compulsion for eternally finding novel methods for its collection. Further, infinite gathering is only matched by a concurrent demand that whatever ways it is used it is economically profitable. Human existence becomes, in this regard, the natural resources for the continuously updating and expanding industry of data exploitation. Each new personal desire and professional aspiration is repackaged as a fresh and inventive monitoring opportunity. What were once seen, therefore, as limitations – surveillance 'gaps' – are transformed into exciting economic and marketing opportunities.

Significantly, this entails empowering individuals to take personal responsible for being 'enterprising' surveillance subjects. This ethos reflects a crucial educating role of the neoliberal state which requires them to implement 'a programme of deliberate intervention by government in order to encourage particular types of entrepreneurial, competitive and commercial behaviour in its citizens'.[69] Refreshed for the digital age, this has evolved into the creation of a culture that makes demands on individuals to identify gaps in surveillance and find innovative means of overcoming them. The app and start-up industry, for instance, represents the merging of a traditional market rationality with the critical democratisation of monitoring. Less explicitly oriented along economic lines are the daily 'hacks' people use to track their behaviour in order to make their lives easier. It is an insidious capitalising upon our digital control, fostering entrepreneurial activities for extending the virtual gaze of our cyber-Big Brother.

Smart Big Brother

The present-day Big Brother is feared not just for being bigger but also smarter. Data technology has made surveillance more predictive, adaptable, 'intelligent' and invasive. This is exacerbated by cultures of self-tracking, enterprising monitoring and data policing. Critical data studies have in turn unpacked how 'data assemblages do work in the world with respect to dataveillance and the erosion of privacy, profiling and social sorting, anticipatory governance, and secondary uses and control creep'.[70] While these efforts are undeniably valuable, it is equally imperative to interrogate the appeal of this digital control and authoritarian regime of totalveillance.

The popular pull of big data transcends its potential economic or social utility. It has become the key to unlocking all of the universe's deepest mysteries. It is a doorway into resolving our most fundamental existential and spiritual queries. Hence, 'Our Ability to capture, warehouse, and understand massive amounts of data is changing science, medicine, business, and technology. As our collection of facts and figures grows, so will the opportunity to answer fundamental questions.'[71] We are on the verge, always moving one data byte closer, to digital enlightenment.

What this uncovers is a crucial element of our contemporary 'big data' fantasy. The digital era presents us with a profound paradox of freedom. We seem to have more choices than ever but less historical agency to shape our personal and collective destinies. This is only compounded by the sheer amount of information we have at our fingertips and little actual power to truly use it to transform our lives or the world. These contradictions, between capabilities and freedom, knowledge and agency, give rise to desires for an omnipotent all-wise person or force that can harness and control big data.

Longed for, in this respect, is a ruler that is all-knowing and as diverse in their powers as the digital realities they are seeking to rule. What is required in this age of disaggregated capitalism is a multitasking autocrat. What is desired is a technological ruler with 'a broad-based capacity extended through society that can act on a variety of inputs to manage emerging knowledge-based technologies while such management is still possible'.[72] Here, the many-tentacled character of contemporary surveillance is turned into a compelling dream of a perfectly automated life. Big data is like an imperceptible manager making sure everything is well ordered and runs smoothly. These longings are exemplified in the popularity of products such as Amazon Echo, which is sold as a type of updated personal planner all the while collecting intimate data about how you eat, sleep, love and play for the real corporate masters.

This growing desire for personal management has been upscaled into a romanticised vision of hi-tech predictive rule. Data-based authoritarianism is founded on its ability to foretell the future – to see through the matrix of a complex virtual world so that it can be more easily navigated and conquered.[73] Crucial to this anticipatory governance is the faith that it can be redirected towards the ends of social and economic justice. In particular,

> Such is also the importance of engagement together with foresight and integration. While changing venues and amplifying within them the still, small voices of folks previously excluded from offering constructive visions of futures may not be complete solutions to our woes in governing technology, they can certainly contribute to bending the long arc of technoscience more toward humane ends.[74]

This explicitly ethical dimension feeds into neoliberal discourses combining monitoring with entrepreneurship. People are attracted to a datafied fantasy of the innovative, successful and in-control entrepreneur. In this respect,

> One secures identity not in 'being' an enterprising subject but in the gap between the subject and the object of desire. Not only does it not matter that the object is unattainable. This lack is central to maintaining desiring. And, as Lacan indicates, if we ever achieve the object of desire, it collapses – it falls apart and is changed inexplicably into a gift of shit.[75]

Ironically, this longing to feel in control through being a perfect digital entrepreneur is precisely what allows elites to exert greater surveillance and control over them.

Importantly, this investment in the ideal of an all-knowing and cyber-wise digital entrepreneur has ominous implications for democracy. It justifies the need for a personal authority who can see through the data noise and decisively govern society. Consequently, the 'Omnipresence of anticipatory governance is felt in the proliferation of focus groups, consensus conferences, Internet surveys, and Wiki and other interactive media – all of which, again intentionally or not, serve to cast doubts on the representativeness of classic democratic institutions like legislatures and elections'.[76] Here lie the seeds of an authoritarian populism that revolves around the dangerous personality cult of the 'CEO politician'.[77] It is a symptom of the more fundamental contemporary desire for a 'smart Big Brother'.

Totalitarianism 4.0

We are entering the age of totalitarianism 4.0. It is marked by a culture of totalveillance, where all aspects of our real and virtual lives are monitored and analysed for greater profit and control. Even more troubling, it is our innovative entrepreneurial spirit that fills existing surveillance gaps. All the while those in power largely escape such monitoring, free from the ever-expanding 'digital eye' of the present-day 'smart Big Brother'. This reflects how the free market is not just politically supported by authoritarian governance but actively shapes its methods and strengthens its rule. Big data has transformed us into 'smart subjects', investing in and working hard for our increasingly total monitoring and exploitation.

8
The Revolution Will Not Be Monitored

In 1970 the late great radical artist Gil Scott-Heron released the incendiary anthem 'the revolution will not be televised'. It was a protest against a mass media that was seeking to co-opt bottom-up struggles for the sake of producing cheap thrills and mass entertainment that could be easily consumed from the privileged comfort of your home. Instead he calls on people to pay attention to what is actually happening in real life – in their cities, communities and workplaces. He sings:

> You will not be able to stay home, brother
> You will not be able to plug in, turn on and drop out
> You will not be able to lose yourself on skag and skip
> Skip out for beer during commercials
> Because the revolution will not be televised

Today there is a new battle cry of freedom emerging – 'the revolution will not be monitored'.

Thus far, this book has highlighted the exploitive and authoritarian aspects of big data and digital technology. Its aim has been to reveal how we have entered an age of 'virtual power' – where the entirety of our concrete, online and even imagined existences are being tracked, controlled and used for economic profit. It encompasses our multiple selves, realities, inner desires, and aspiring futures. Further, it reveals the troubling digital merging of surveillance and market logics, as people increasingly become enterprising datafied subjects who must innovatively find ways to personally fill the monitoring gaps of economic and political elites. This form of data empowerment serves to enable and strengthen a neoliberal culture of totalveillance and the rise of an even more powerful contemporary smart Big Brother. Critically, this same monitoring culture is not extended to elites nor to the capitalist system itself.

What we have uncovered is the full extent of the present-day paradox of (post-)modern virtual power. The bigger the data the smaller our ability to monitor, hold accountable and ultimately transform the status quo. Instead the fragmented and disaggregated nature of contemporary society leads to greater demands for systematic surveillance and personalised monitoring regimes. Moreover, this dialectic of digital control is reinforced by the immaterial labour we constantly expand as data explorers and entrepreneurs. Consequently, we are not only made complicit in our own economic mining as an infinitely renewable data resource, but also in our growing domination by the spread of 'totalitarianism 4.0'.

Yet all hope is not lost. The last several years have been rocked by a worldwide resurgence of 'anti-establishment' politics from both the left and the right. These political earthquakes are rightly linked to broad swathes of the population feeling 'left behind' by globalisation. However, they were reacting against being plugged into a digitised free market system that was outside of their control and was secretly being used against them. It is not a coincidence that these movements have turned the tables on those in power by exploiting social media and open-source information for resisting elite rule.

While it is easy and legitimate to bemoan the reactionary results of these data-based popular revolts, they also reveal a profound present-day political truth. A key part of any current struggle for social change must prominently include reversing the oppressive paradox of digital control. The celebration of whistleblowers such as Edward Snowden, and to a lesser and much more controversial extent Julian Assange, reflects the popular clamouring for redirecting such invasive surveillance against those in power.[1] It also reveals the radical reconfiguration of this present capitalist dialectic: the more disaggregated the power and methods of these elites become, the more they need to be tracked, critically analysed and made publicly accountable. It is only in doing so that 'virtual power' can be a liberating force that allows us to explore the full potentialities of our possible selves, realities and imagined futures.

Reversing the Paradox

A crucial question for our big data times is how can monitoring be used to benefit the many and not the few? Our increasingly connected world is simultaneously over and under-monitored. The majority are under almost constant surveillance and scrutiny. The powerful use this same technology to evade responsibility and hide their actions from public view. The question is not whether we are being watched but rather who is being watched and why?

An important legitimation of such digital control is the need for constant oversight. Predictive analysis can tell you whether an employee or student will fail or succeed. Yet this always assumes the perspective of authority, even when done with the best of intentions. It presupposes a need for management, for containing and shaping the conduct of those being analysed from a position of power. What is missing is a sense of 'undersight' based on 'sousveillant' practices of everyday monitoring of authority that can become 'a potentially effective political force that can now challenge and balance the hypocrisy and corruption that is otherwise inherent in a surveillance-only society (i.e. a society that has only oversight without undersight)'.[2] It is necessary to reorient monitoring from the bottom up – to start from the point of view of how those with power and authority can be identified and monitored, and how to prevent elites from engaging in mass surveillance.

Key to this power shift is the democratisation of big data. Specifically, the transformation of data-driven into data-based democracies. In particular,

> data mining and analytics need to be democratised in three ways: they should be subject to greater public supervision and regulation, available and accessible to all, and used to create not simply known but reflexive, active and knowing publics. We therefore imagine conditions in which data mining is not just used as a way to know publics, but can become a means for publics to know themselves.[3]

Such democratisation would also serve as the means to better understand macro and micro power relations, to allow us to learn not just more about ourselves but about how we are exploited and controlled. Activists and policymakers are already drawing on big data to track global human trafficking – the next step is labour exploitation internationally and the potential for creating a more liberated world.

This critical reprogramming entails the shift away from big data to 'reflexive' data. This involves the adoption of techniques meant to invoke digital questioning rather than mere consumption and conformity. Researchers are already proposing the notion of 'digital orality' as a new form of meaningful storytelling using big data.[4] More broadly, this would permit the retelling of the development of computing and data, diverting it away from current capitalist narratives. The heroic and visionary story of digital robber barons such as Bill Gates, Steve Jobs and Mark Zuckerberg would be replaced with ones that reveal the collective, collaborative and democratic impulses driving these developments.[5]

It also means deploying data for specific political, ecological and humanitarian purposes.[6] These efforts would fundamentally start to redirect virtual power. It would reverse the paradox of our digital world – monitoring and challenging power relations instead of providing them with covert and overt support. Consequently, data could become that which empowers our personal and social imagination as opposed to the force seeking to constrain and control it.

Remonitoring Power

We must do more than simply repurpose big data, however. Instead, it is vitally important to redeploy it to remonitor power – specifically, to put current power relations under the digital microscope to examine their impact upon our everyday lives and across our lifespans. It means recognising how it affects our digital virtual selves, narrowing our online possibilities and decreasing our freedom in real life. Doing so entails the fostering of alternative monitoring regimes of power.

In the short term it means radicalising existing ideas of 'digital citizenship'. To this effect,

> Digital Civics is an emerging cross-disciplinary area of research that [is] seeking to understand the role that digital technologies can play in supporting relational models of service provision, organization and citizen empowerment. In particular, how digital technologies can scaffold a move from transactional to relational service models, and the potential of such models to reconfigure power relations between citizens, communities and the state.[7]

While these civic technologies still largely conform to liberal values of representational democracy, and market assumptions of service provisions, they hold the promise to expand the potential of twenty-first-century politics and emancipation. Technologies such as blockchain are currently transforming 'money, business, and the world' – gradually discrediting the state's monopoly of authority when it comes to money.[8] Also emerging is the 'rise of the mediating citizen' able to use social media techniques such as 'crowdsourcing' to directly influence political policy.[9]

At a smaller scale, it entails finding ways to employ these digital advances to produce novel more progressive 'data publics'.[10] Rather than simply using social media to create online communities or better connect physical communities, it can be critically exploited to question these technologies and their role in producing more egalitarian social relations. Thus by drawing on collaborative user-led data perspectives and practices such as 'data walks' and 'writing free-for-alls', individuals can start discussions as to what should constitute data and how should it be deployed for their shared benefits.[11] More broadly, it helps to redefine the definition and scope of public data empowerment. Present-day, 'open data movements' can 'rearticulate notions of democracy, participation, and journalism by applying practices and values from open source culture to the creation and use of data'.[12] In particular, it encourages activists to share raw data and push for an 'open source model of participation to political participation', and to view their journalism as a form of critical digital praxis in which they experiment with tuning their ideas into realities.

Big data, in this respect must be ideologically and concretely delinked from its roots in capitalism. Its revolutionary potential is held back by its free market origins and biases. However, as the social critic Pankaj Mehta recently proclaimed in the left-wing publication *Jacobin*:

> Big data, like all technology, is imbued within social relations. Despite the rhetoric of its boosters and detractors, there is nothing inherently progressive or draconian about big data. Like all technology, its uses reflect the values of the society we live in. Under our present system, the military and government use big data to suppress populations and spy on civilians. Corporations use it to boost profits, increase productivity, and extend the process of commodification ever

> deeper into our lives. But data and statistical algorithms don't produce these outcomes – capitalism does. To realize the potentially amazing benefits of big data, we must fight against the undemocratic forces that seek to turn it into a tool of commodification and oppression. Big data is here to stay. The question, as always under capitalism, is who will control it and who will reap the benefits.[13]

The real task, then, is whether we can decapitalise digital technology and socialise it for more progressive, just and emancipatory ends?

That involves transforming the threat of totalveillance into the exciting potentialities of 'powervelliance'. On the one hand, that entails deploying digital surveillance methods to understand existing power relations and inequalities as thoroughly and systematically as possible. On the other, it invites the use of these same technologies for developing alternative non-capitalist forms of daily existence. It means asking 'Instead of (designing for) desiring commodities and self-improvement, how can we (design for) alternative desires and ways of feeling with/through data?'[14] Researchers, for instance, are using biosensing technology to design data-driven visuals that 'resists quantification and centralization of its data; instead, it invites situated in-the-moment curiosity and a different way of experiencing everyday surroundings. It carves a subversive path through the urban datascape of the optimized city. Can it give us license to desire, to feel, and to crawl rather than to measure up to over-fitted expectations?'[15] This is a compelling example of how the digital can remonitor and in fact reconfigure contemporary power.

Infinite Data

At stake is how the so-called 'data revolution' can expand rather than restrict social possibility. Theoretically, the theorist Michel

Foucault helps to rethink power away from strict forms of coercion and instead in terms of 'fields of possibility'. He states that 'Power exists only when it is put into action, even if, of course, it is integrated into a disparate field of possibilities brought to bear upon permanent structures.'[16] This shaping of the possible is also strategic, as he highlights by asking: 'Rather than seeking the permanence of themes, images, and opinions through time, rather than retracing the dialectic of their conflicts in order to individualize groups of statements, could one not rather mark out the dispersion of the points of choice, and define prior to any option, to any thematic preference, a field of strategic possibilities?'[17] Rephrasing this query for present times, in what way is big data narrowing and enlarging our own personal and collective possibilities?

In this spirit, social media can be used as a prime force for revealing the contemporary social and political limits of big data. Blogs, for instance, can be drawn upon to raise questions as to whose voices and perspectives are not being heard and why.[18] It also permits the emergence of 'messy information' into the public sphere.[19] These techniques can be especially effective in contentious and conflictual contexts such as war zones. The 'warblogs' of Iraqi women, to this end,

> can be understood as practices of the self which provide a glimpse into a number of intersecting, competing and conflicting fields of possibility in Iraq ... Fields of possibility are more or less discursively constructed social spaces and are highly mobile in their relation to the unfolding self. In online fields we are not just faced with chat rooms or rants, but with fragments of the self and windows into life-worlds. We are able to see the self in the making: who we are, what we thought at any given time in relation to other people and places. I argue for the emergence of a digital self through online spaces.[20]

Instead of seeking to provide greater clarity or predict their behaviour, these raw data illuminated the complexity of 'limits of freedom' experienced by these subjects.

In turn, these practices can give rise to 'bottom-up' forms of data-driven politics. This implies much more than 'looking upwards' and surveilling prevailing power. It activates, by contrast, the agency of individuals and communities to transcend the limitations placed upon them by 'algorithmic power',[21] specifically by revolving community building and the deployment of analytics around the need to expand rather than restrict social voice. These practices are captured in the emergence of radical 'cyber-activism'. Female bloggers during the Arab uprisings were known as the 'twitterati' for their large number of followers and their daily role in inspiring these insurrections.[22]

A full-scale 'digital rebellion' is beginning to emerge, in which social media and big data are reconfiguring the very possibilities of social movements and change. Across the world they are allowing activists and everyday citizens to network for spreading alternative ideologies and organising direct and subversive actions that are challenging the once thought to be permanent status quo.[23] These cyber-uprisings are still obviously incipient, and in many ways as successful from the reactionary right as they are the progressive left. However, they represent the beginnings of 'cyber-propelled revolutions' based on 'computer-mediated communication'.[24] Solidarity will be built and sustained going forward as much online as it will be offline.

Yet this revolutionary turn is not and cannot be confined to simplify radical movements for social transformation. They must also direct their attention to altering the very perception of data itself. In its current form, it is that which collects information about us so that it can control our present and predict our futures. It is concerned not so much with possibility but enlarging itself, making itself 'bigger', so that it can know more about who we are and will be. Conversely, a radical view would be to promote the potential for 'infinite data', the championing of digital techniques and analytics for highlighting our social possibilities rather than accurately assessing our personal and collective limitations.

Virtually Freeing Ourselves

This book has introduced the concept of virtual power. In the new millennium, the scope of hegemony and control has dramatically expanded. Power is no longer content to remain trapped in real life. To be confined by actual events and concrete relationships. Nor does it want to necessarily restrict what is socially possible, as it has in the past. Rather, its focus is on discovering innovative ways for mining people's virtual imaginations and exploiting their diverse online and offline existences.

To counter this virtual assault, it is necessary to engage in concerted and radical speculation. It is often lamented that the present era is being destroyed by 'fake news'. Certainly, social media and cyber-interactions have revealed the biases lurking behind every 'fact' and perspectival slant underpinning every 'truth'. What was once a discussion over the accuracy of news reporting has morphed into 'whose fictional reality do you believe the most?' While this undoubtedly poses fresh challenges for activist and citizens alike, a politics of real-time fact checking is simply not enough. Instead, it is crucial to draw on this virtuality in order to promote fresh visions of a better future. Leading critical thinkers such as Adrienne Maree Brown and Walidah Imarisha are attempting to perform just such a radical intervention in their notion of 'visionary fiction'. They declare that 'Whenever we try to envision a world without war, without violence, without prisons, without capitalism, we are engaging in an exercise of speculative fiction. Organizers and activists struggle tirelessly to create and envision another world, or many other worlds, just as science fiction does.'[25]

While visionary fiction is specifically targeting the subversive implications of genres like science fiction, it points the way towards the potential for virtual revolutions. There is a growing interest in the function of simulations – in the form of online games and virtual reality – for expanding our personal and cultural horizons. Through the simple act of putting on a headset people can now immerse themselves in any culture around the world, exploring the far reaches of space as well as the distant past. They can also use internal online networks for 'dreaming' about the possibilities of democracy in their workplace[26] or society at large.[27] Yet they can also provide us with much more far-reaching and revolutionary dreams. These could include immersing ourselves in a 'post-capitalist future' or a coming 'world without prisons'. Virtual technology can therefore make what was once merely hypothetical and abstractly desirable into lucid, present-day realities.

These hypothetical revolutions have the power to make what could be socially disruptive technologies into empowering and exciting ones. The coming prospect of AI, robots and the further growth of big data fills a growing number of people with dread. Instead of conjuring up utopian scenarios of progress, they are more likely to shudder with fear over a techno-dystopian tomorrow filled with mass unemployment, environmental ruin and social disconnection. Yet digital technologies are already reframing these disruptions into viable market opportunities. Practices such as 'life logging' turn fear over the future into a daily regime of 'self-tracking' personal progress and self-improvement.[28] Fundamentally, big data is being used as a tool for 'managing risks for disruptive technologies', providing investors with a sense of long-term security and everyday people with a feeling of existential safety.[29] Similarly, these same advances can be turned into hi-tech resources for reimagining and progressively realising a future that disrupts our neoliberal status quo for a more egalitarian, emancipated and just world.

Most explicitly, this would involve drawing on virtual reality to promote a different type of social order. This entails recognising the threat posed by 'virtual capitalism', in which our agency, labour and critical imagination is exchanged for the fleeting pleasures of virtual entertainment:

> Here's how virtual capitalism works: NKK, a Japanese steel company with a failing shipyard, converts the shipyard into a facility to produce simulated domed beaches complete with wave-making machines and surfing contests. The selling point is that nothing unpleasant, uncomfortable, or inconvenient happens at these beaches: the last man's paradise. Virtualization in the name of exchange value is the formula for the transition from industrial capitalism to virtual capitalism.[30]

Within our daily experiences, we are currently witnessing the transition from 'socialism' to market-friendly forms of 'sociality online', in which 'likes' and 'mentions' serve as 'digital gifts' that represent both mutual recognition and the tacit support of huge corporate platforms.[31]

Nevertheless, this dystopian vision of our virtual futures is by no means predetermined. It is not the only course our digital destinies can take. It is crucial, therefore, to engage in a new virtual struggle for freedom and equality. More precisely, to 'take back' control of our imagination and the fictions that shape our experiences and dominate our lives. Doing so means deobjectifying big data capitalism – stripping off its veneer of scientific objectivity and empiricism. As noted critical thinker William Davies argues in the introduction to his appropriately named edited collection *Economic Science-Fictions*, 'Far from being a system liberated from fictions, capitalism should be seen as a system that liberates fictions to rule over the social … it must be stressed here that fictions are not necessarily falsehoods or lies, far from it. Economic and social fictions elude empiricism, since they are never given in experience, they are what structures experience.'[32]

The pervasive surveillance regimes and invasive monitoring culture associated with the big data fictions structuring our own times demand a resistance politics that writes its own liberating fictions – ones that highlight the possibilities of being unbound by the ideological chains of the free market and where people are largely unmonitored, while the system that governs them (as well as the elites trying to exploit them) remain firmly within public view and under our control.

Digital Revolutions

The danger of these virtual insurgencies and radical futures is that they will never come to fruition 'in real life'. Just like great internet friends who never meet, or when they do it is not as they imagine, digital insurrections risk losing steam when there is an attempt to make them physical. In much the same way, online social transformations have a strange way of never becoming an offline reality. In order to avoid this fate, to sidestep being tossed into the dustbin of our virtual histories, it is crucial to translate and integrate these cyber-revolutions into our daily physical existence.

It is here that the virtual, the concrete and the political all converge and mingle. Dramatic change is not just being brought to us by big data and digital technologies. It is also being driven by nano-technologies, robotics and genetic engineering. These advances will utterly alter what it means to be 'human' and 'alive'. Nevertheless, they bring with them refreshed political concerns and struggles over how democratic, safe, accessible and free these innovations to our very existence will be. In his landmark 2005 book *Cyborg Citizen*, theorist James Hughes prophesied that 'becoming more than human can improve all our lives, but only new forms of trans-human citizenship and democracy can make us freer, more equal and more united'.[33] These hopes and fears are shared by the rise of the 'cyber-citizen', with worries that online voting and network-based civic discussion would be more repressive than politically empowering.[34] Moreover, the precise virtuality of these encounters, it has been suggested, is inherently undemocratic, as 'the ideal of democratic politics relies on the notion of the "commons" as a real space for political activity, debate, and exchange. Virtual space cannot provide a substitute. Democratic politics must have as its premises real bodies, confronting real problems, in real space'.[35]

While these concerns are certainly valid and still timely, they risk essentialising these technologies. Rather, it is how they are used and for what purpose that is most crucial. The employment of information technology is a prime example, as there are 'important differences' in how it 'is used in military and social-movement cultures'. The former deploys it according to a 'security-police mode for quantifying and controlling social space, in order to meet low-intensity, counterinsurgency, and regime-maintenance goals (or for recruitment and public relations)'. By contrast, 'For social-movement cultures, such as secular Egyptian revolutionaries, 15M (Los Indignados), and Idle No More, social media is an integral part of life; it is context. Unlike these horizontalist movements, military institutions are based on a hierarchical structure that precludes social media from becoming part of their organizational and decision-making culture.'[36] Just as significantly, 'The differences are more than a matter of how the affordances of information technologies match with the different technocultures. Horizontalist social movements incorporate new information technologies into their praxis as self-control, while militaries seek to subsume them into the existing hierarchical control paradigms.'[37]

What is critical, in this respect, is the ability to use digital skills to reconfigure social relations in both big and small ways. The image of the hacker within popular culture largely resides around an almost nihilistic figure of cultural subversion. The computer genius holed up alone in their room wreaking havoc for its own sake. Put in a more positive light, they are the 'whizzes' in the blockbuster movies and TV shows who can break into any system and find out any background information so that the hero can succeed. These are hardly revolutionary notions. A noted departure is the show 'Mr. Robot', where the hacker protagonists engage in explicit revolutionary activity, to mixed effect. Beyond entertainment, 'civic' hacking culture holds the potential to redirect big data to create more sustainable and fair 'smart cities' that challenge market-based models of urban development.[38] Such 'hacktivism' can fuel real, global revolutionary movements.[39] In turn, we see the broader need to recognise that people value data differently and that it is imperative to deploy data in a way that empowers people as part of their 'inter-operability' – their everyday routines and practices.[40]

What we require are new forms of 'radical intelligence' that can break through the narrow perspectives of big data capitalism and its digital control. The aim, in this respect,

> is to rediscover the role of error, trauma and catastrophe in the design of intelligent machines and the theory of augmented cognition. These are timely and urgent issues: the media hype of singularity occurring for artificial intelligence appears just to fodder a pedestrian catastrophism without providing a basic epistemic model to frame such an 'intelligence explosion'.[41]

Rather than focus on prediction or the 'objective' truth of individuals, what should be valued instead is how data can 'queer' our conventional understandings. Hence, it is well established that self-tracking apps, such as those related to sexual and reproductive processes, 'work to perpetuate normative stereotypes and assumptions about women and men as sexual and reproductive subjects'.[42] Yet if they were repositioned to reveal the sheer diversity of our preferences in this most intimate and personal of areas, this could trouble dominant gender assumptions and open the way for us to embrace new possibilities and different cultural practices.

Consequently, data would be responsible for producing a new and more radical set of social desires. Currently identity is linked to becoming whole through optimising ourselves via digital monitoring and tracking our 'deeper data' across our lifetime. Yet if we reoriented ourselves to a revolutionary data project of self and social creation, where the goal is not to be made whole or achieve perfection but to explore our virtual possibilities, then data would undergo a 'radical democratic' transformation away from the exploitative capitalist roots. This transformation echoes the renowned psychoanalytic political theorist Yannis Stavrakakis's view of radical democracy, 'not merely as an aggregate of different interests or a constitutional structure based on human and political rights, but – above all else – as an institutionalisation of lack and antagonism, as the possibility of instituting a sustainable and interminable questioning which permits the reflexive self-creation of society'.[43] When big data comes to represent our existential freedom to transform our society in unexpected and progressive ways, then it will have moved from a tool of empowering control into a force of unpredictable revolution.

The Revolution Will Not Be Monitored

We are rapidly entering the age of virtual power. While big data is supposedly meant to make us more predictable, it has in fact revealed the infinite possibilities we have still have for reimagining our present and future. Anything that we can virtually imagine can become an immersive reality, first digitally and then in real life. The 'objectivity' of capitalism, the free market and oligarchy need not be the end of our histories. Instead they can be the very platforms that we jump off from into more creative and egalitarian worlds to come.

Yet it is for this very precise reason that big data is being enhanced as a tool for controlling not only our present actions but also our not-yet dreamed of potentialities. The limitlessness of our current digital condition produces a concurrent need to guide, narrow and direct it towards certain ideological and profitable ends. Further, the very gaps in the system demand that for us to be totally surveilled we must be transformed into enterprising, self-monitoring subjects. The empowerment found in using our data to explore virtual realities is exactly that which paradoxically leaves elites and the exploitive system supporting them unwatched and publicly unaccounted for. The fact that the impossible is now increasingly possible means it must be strictly monitored in case it disrupts the status quo.

These big data times, though, do not have to be ruled by the desires and whims of surveillance and business. It does not have to be twentieth-century capitalism and control rebooted for a new millennium. Our information does not have to be our most valuable resource, to be colonised and mined by the invasive forces of seen and unseen hi-tech corporate overlords. We do not have to live our diverse existences under the watchful electronic eye of a 'smart Big Brother'. Instead we can forge ahead with 'infinite data' rather than big or deeper data, exchanging digital control for data-based potentialities, virtual power for virtual possibilities. The future is ours to log on to and explore, giving voice to the once unheard, shedding light on our complicated presents and our exciting tomorrows, because the revolution will not be monitored.

Notes

Chapter 1

1. Greenfield, P. (2018). 'The Cambridge Analytica Files: The Story So Far'. *The Guardian*, 26 March.
2. Devine, C., O'Sullivan, D. and Griffin, D. (2018). 'How Steve Bannon Used Cambridge Analytica to Further His Alt-right Vision for America'. CNN, 16 May.
3. Cadwalladr, C. and Graham-Harrison, E. (2018). 'How Cambridge Analytica Turned Facebook "Likes" into a Lucrative Political Tool'. *The Guardian*, 17 March.
4. Bartlett, J. (2018). 'Big Data is Watching You – and it Wants Your Vote'. *The Spectator*, 24 March.
5. Cadwalladr, C. (2017). 'The Great British Brexit Robbery: How Our Democracy Was Hijacked'. *The Guardian*, 7 May.
6. Irani, D. (2017). 'Why is Ex-adman Nigel Oakes Being Hailed As the "007" of Big Data?'. *Economic Times*, 29 March.
7. Persily, N. (2017). 'Can Democracy Survive the Internet?'. *Journal of Democracy*, 28(2): 63.
8. González, R. J. (2017). 'Hacking the Citizenry?: Personality Profiling, Big Data, and the Election of Donald Trump'. *Anthropology Today*, 33(3): 9–12.
9. Fuller, M. and Goffey, A. (2012). *Evil Media*. MIT Press: 5. Also see Grusin, R. A. (2017). 'Donald Trump's Evil Mediation'. *Theory and Event*, 20(1): 86–99.
10. Laterza, V. (2018). 'Cambridge Analytica, Independent Research and the National Interest'. *Anthropology Today*, 34(3): 1–2.
11. Gross, M. (2018). 'Watching Two Billion People'. *Current Biology*, 9(7): 527–30.
12. Davies, W. (2018) 'Short Cuts'. *London Review of Books*, 40(7): 20.
13. Williamson, B. (2018). 'Why Education is Embracing Facebook-style Personality Profiling for Schoolchildren'. *The Conversation*, 29 March 2018.

14. See Baldwin-Philippi, J. (2017). 'The Myths of Data-driven Campaigning'. *Political Communication*, 34(4): 627–33.
15. Beer, D. (2018). 'Envisioning the Power of Data Analytics'. *Information, Communication & Society*, 21(3): 465–79.
16. Beer, D. (2017). 'Data-led Politics: Do Analytics Have the Power that we Are Led to Believe?'. *British Politics and Policy at LSE*, 3 March 2017.
17. Fuchs, C. and Sevignani, S. (2013). 'What is Digital Labour? What is Digital Work? What's their Difference? and Why Do These Questions Matter for Understanding Social Media?'. *Triple C*, 11(2): 288.
18. Lupton, D. (2016). 'The Diverse Domains of Quantified Selves: Self-tracking Modes and Dataveillance'. *Economy and Society*, 45(1): 101–22.
19. Zuboff, S. (2018). *Master or Slave? The Fight for the Soul of Our Information Civilisation*. Profile Books: back cover.
20. White House (2014). 'Big Data: Seizing Opportunities, Preserving Values'. Report for the President, Executive Office of the President, Washington, DC.
21. Cohen, J. E. (2015). 'The Surveillance-innovation Complex'. In Barney, D., Coleman, G., Ross, C., et al. (eds): *The Participatory Condition*. University of Minnesota Press.
22. Roberts, L. (2016). 'Deep Mapping and Spatial Anthropology'. *Open Access Humanities*, 5(1): 5.
23. Zuboff, S. (2015). 'Big Other: Surveillance Capitalism and the Prospects of an Information Civilization'. *Journal of Information Technology*, 30(1): 75.
24. Lin, Y. W. (2018). '#Deletefacebook is Still Feeding the Beast – But There Are Ways to Overcome Surveillance Capitalism'. *The Conversation*, 26 March.
25. See Peper, E. and Harvey, R. (2018). 'Digital Addiction: Increased Loneliness, Anxiety, and Depression'. *NeuroRegulation*, 5(1): 3; Alrobai, A., McAlaney, J., Dogan, H., Phalp, K. and Ali, R. (2016). 'Exploring the Requirements and Design of Persuasive Intervention Technology to Combat Digital Addiction'. In Bogdan, C. et al. (eds): *Human-Centered and Error-Resilient Systems Development*. Springer.
26. Schneier, B. (2015). *Data and Goliath: The Hidden Battles to Collect Your Data and Control Your World*. W. W. Norton and Company: 238.

27. Lomborg, S. and Frandsen, K. (2016). 'Self-tracking As Communication'. *Information, Communication & Society*, 19(7): 1015.
28. Ganesh, S. (2016). 'Digital Age/Managing Surveillance: Surveillant Individualism in an Era of Relentless Visibility'. *International Journal of Communication*, 10 (14): 166.
29. Sharon, T. and Zandbergen, D. (2017). 'From Data Fetishism to Quantifying Selves: Self-tracking Practices and the Other Values of Data'. *New Media & Society*, 19(11): 1695–709.
30. Powell, J. A. (1996). 'The Multiple Self: Exploring Between and Beyond Modernity and Postmodernity'. *Minnesota Law Review*, 81: 1484.
31. Balsamo, A. (1995). 'Forms of Technological Embodiment: Reading the Body in Contemporary Culture'. *Body & Society*, 1(3–4): 215.
32. Kafai, Y. B., Fields, D. A. and Cook, M. S. (2010). 'Your Second Selves: Player-designed Avatars'. *Games and Culture*, 5(1): 23–42. Also see Fox, J. and Ahn, S. J. (2013). 'Avatars: Portraying, Exploring, and Changing Online and Offline Identities'. In Luppicini, R. (ed.): *Handbook of Research on Technoself: Identity in a Technological Society*. IGI Global.
33. Boellstorff, T. (2015). *Coming of Age in Second Life: An Anthropologist Explores the Virtually Human*. Princeton University Press: 57.
34. Martey, R. M. and Consalvo, M. (2011). 'Performing the Looking-glass Self: Avatar Appearance and Group Identity in Second Life'. *Popular Communication*, 9(3): 165.
35. Carton, F., Brezillon, P. and Feller, J. (2016). 'Digital Selves and Decision-making Contexts: Towards a Research Agenda'. *Journal of Decision Systems*, 25(Sup. 1): 96.
36. Goffman, E. (1959). *The Presentation of Self in Everyday Life*. Anchor Books. See also Kerrigan, F. and Hart, A. (2016). 'Theorising Digital Personhood: A Dramaturgical Approach'. *Journal of Marketing Management*, 32(17–18): 1701–21.
37. Hicks, T. (2010). 'Understanding and Creating Your Digital Self'. *Psychology Today*, 23 August.
38. Marvin, C. (1990). *When Old Technologies Were New: Thinking About Electric Communication in the Late Nineteenth Century*. Oxford University Press.

39. Boon, S. and Sinclair, C. (2009). 'A World I Don't Inhabit: Disquiet and Identity in Second Life and Facebook'. *Educational Media International*, 46(2): 99–110. Baym, N. K. (2015). *Personal Connections in the Digital Age*. John Wiley and Sons.
40. Trub, L. (2017). 'A Portrait of the Self in the Digital Age: Attachment, Splitting, and Self-concealment in Online and Offline Self-presentation'. *Psychoanalytic Psychology*, 34(1): 78.
41. Cheney-Lippold, J. (2017). *We Are Data: Algorithms and the Making of Our Digital Selves*. NYU Press.
42. Elwell, J. S. (2014). 'The Transmediated Self: Life Between the Digital and the Analog'. *Convergence*, 20(2): 233–49.
43. Ahn, S. J. G., Phua, J. and Shan, Y. (2017). 'Self-endorsing in Digital Advertisements: Using Virtual Selves to Persuade Physical Selves'. *Computers in Human Behavior*, 71: 110–21.
44. Barber, M. V. (2018). 'The Risk of Privately Owned Public Digital Place'. *Risk*, 11 April.
45. Bevir, M. (2011). 'Governance and Governmentality After Neoliberalism'. *Policy & Politics*, 39(4): 457–71.
46. Ong, A. (2007). 'Neoliberalism As a Mobile Technology'. *Transactions of the Institute of British Geographers*, 32(1): 3–8.
47. Giroux, H. A. (2018). *Terror of Neoliberalism: Authoritarianism and the Eclipse of Democracy*. Routledge.
48. Ball, K. (2017). 'All Consuming Surveillance: Surveillance As Marketplace Icon'. *Consumption Markets & Culture*, 20(2): 95–100.
49. Winsborough, D., Lovric, D. and Chamorro-Premuzic, T. (2016). 'Personality, Privacy and Our Digital Selves'. *The Guardian*, 18 July.
50. Hobbs, P. (2017). '"You Willingly Tie Yourself to These Leashes": Neoliberalism, Neoliberal Rationality, and the Corporate Workplace in Dave Eggers' *The Circle*'. *Dandelion: Postgraduate Arts Journal and Research Network*, 8(1): 1.
51. Baym, N. K. (2015). *Personal Connections in the Digital Age*. John Wiley and Sons.
52. Sjöberg, M., Chen, H. H., Floréen, P., Koskela, M., Kuikkaniemi, K., Lehtiniemi, T. and Peltonen, J. (2016). 'Digital Me: Controlling and Making Sense of My Digital Footprint'. In *International Workshop on Symbiotic Interaction*. Springer.
53. Haimson, O. L., Brubaker, J. R., Dombrowski, L. and Hayes, G. R. (2015). 'Disclosure, Stress, and Support During Gender Transition on Facebook'. In *Proceedings of the 18th ACM Conference on*

Computer Supported Cooperative Work & Social Computing. ACM. Also see Haimson, O. L., Brubaker, J. R., Dombrowski, L. and Hayes, G. R. (2016). 'Digital Footprints and Changing Networks During Online Identity Transitions'. In *Proceedings of the 2016 CHI Conference on Human Factors in Computing Systems*. ACM.

54. Braidotti, R. (2011). *Nomadic Theory: The Portable Rosi Braidotti*. Columbia University Press.
55. See especially Yaffe, D. S. (1973). 'The Marxian Theory of Crisis, Capital and the State'. *Economy and Society*, 2(2): 186–232; Mattick, P. (1981). *Economic Crisis and Crisis Theory*. Merlin Press; Clarke, S. (1994). *Marx's Theory of Crisis*. Macmillan.
56. See especially Dore, R., Lazonick, W. and O'Sullivan, M. (1999). 'Varieties of Capitalism in the Twentieth Century'. *Oxford Review of Economic Policy*, 15(4): 102–20.
57. See Bloom, P. and Rhodes, C. (2018). *The CEO Society: The Corporate Takeover of Everyday Life*. Zed Books.
58. See especially Bonefeld, W. (1992). 'Social Constitution and the Form of the Capitalist State'. In W. Bonefeld, R. Gunn and K. Psychopedis (eds): *Open Marxism Vol. I. History and Dialictics*. Pluto.
59. Bell, P., and H. Cleaver. (1982). 'Marx's Crisis Theory as a Theory of Class Struggle'. Research.
60. See Wolfe, A. (1977). *The Limits of Legitimacy: Political Contradictions of Contemporary Capitalism*. Free Press.
61. Bell, D. and Bell, D. (1976). *The Cultural Contradictions of Capitalism, Vol. 20*. Basic Books.
62. O'Connor, J. (1991). 'On the Two Contradictions of Capitalism'. *Capitalism Nature Socialism*, 2(3): 107–9. In particular O'Connor speaks of 'two contradictions' – the first the more conventional 'rate of exploitation', and the second the size and value of consumption alongside the various costs and social externalities for meeting these demands. Both contradictions importantly require an ever-evolving and complex form of technical accounting as well as social accountability.
63. Anderson, B. (1983). *Imagined Communities: Reflections on the Origins and Spread*. Verso.
64. Harvey, David. (2007). *A Brief History of Neoliberalism*. Oxford University.
65. Ibid., 36.

66. Hall, S. (2011). 'The March of the Neoliberals'. *The Guardian*, 12 September.
67. See especially Ouellette, L. (2008). '"Take Responsibility for Yourself": Judge Judy and the Neoliberal Citizen'. *Feminist Television Criticism: A Reader*, 2: 139–53; Room, R. (2011). 'Addiction and Personal Responsibility As Solutions to the Contradictions of Neoliberal Consumerism'. *Critical Public Health*, 21(2): 141–51; Trudeau, D. and Cope, M. (2003). 'Labor and Housing Markets As Public Spaces: "Personal Responsibility" and the Contradictions of Welfare-reform Policies'. *Environment and Planning A*, 35(5): 779–98; Fraser, N. (1993). 'Clintonism, Welfare, and the Antisocial Wage: The Emergence of a Neoliberal Political Imaginary'. *Rethinking Marxism*, 6(1): 9–23.
68. Bloom, P. (2016). *Authoritarian Capitalism in the Age of Globalization*. Edward Elgar Publishing.
69. Demmers, J., Fernandez E. J. A., and Hogenboom, B. (eds) (2004). *Good Governance in the Era of Global Neoliberalism: Conflict and Depolitisation in Latin America, Eastern Europe, Asia, and Africa*. Taylor and Francis.
70. See Jamali, D. R., El Dirani, A. M. and Harwood, I. A. (2014). 'Exploring Human Resource Management Roles in Corporate Social Responsibility: The CSR-HRM Cocreation Model'. *Business Ethics: A European Review*, 24(2): 125–43; Haiven, M. (2014). *Cultures of Financialization: Fictitious Capital in Popular Culture and Everyday Life*. Palgrave Macmillan; Hawkins, D. E. (2006). *Corporate Social Responsibility: Balancing Tomorrow's Sustainability and Today's Profitability*. Palgrave Macmillan.
71. Feher, K. (2016). 'Digital Identity: the Transparency of the Self'. In *Applied Psychology: Proceedings of the 2015 Asian Congress of Applied Psychology*: 132.
72. Schawbel, D. (2009). *Me 2.0: Build a Powerful Brand to Achieve Career Success*. Kaplan Books; Khedher, M. (2014). 'Personal Branding Phenomenon'. *International Journal of Information, Business and Management*, 6(2): 29.

Chapter 2

1. See Bloodsworth, J. (2018). *Hired: Six Months Undercover in Low Wage Britain*. Atlantic Books.

2. Willis, J. (2016) '7 Ways Amazon Uses Big Data to Stalk You (AMZN)'. Investopedia.com, 7 September.
3. Shephard, A. (2017). 'Is Amazon Too Big to Tax?'. *New Republic*, 1 March.
4. See Menzies, H. (1997). 'Telework, Shadow Work: The Privatization of Work in the New Digital Economy'. *Studies in Political Economy*, 53(1): 103–23.
5. Fullerton, J. (2018). 'Suicide at Chinese Iphone Factory Reignites Concern Over Working Conditions'. *The Telegraph*, 7 January.
6. Merchant, B. (2017) 'Life and Death in Apple's Forbidden City'. *The Guardian*, 18 June.
7. See for instance Chu, J. S. and Davis, G. F. (2016). 'Who Killed the Inner Circle? The Decline of the American Corporate Interlock Network'. *American Journal of Sociology*, 122(3): 714–54; Houston, R. and Ferris, S. (2015). 'Does the Revolving Door Swing Both Ways? The Value of Political Connections to US Firms'. *Managerial Finance*, 41(10): 1002–31.
8. Fiss, P. (2016) 'A Short History of Golden Parachutes'. *Harvard Business Review*, 3 October.
9. Griffin, E. (2017). 'A Secret Job Board Opens to the Masses, Sort Of'. *Wired*, 14 September.
10. See Davies, W. (2015). *The Happiness Industry: How the Government and Big Business Sold Us Well-being*. Verso Books.
11. Neff, G. and Nafus, D. (2016). *The Self-Tracking*. MIT Press: 37.
12. Moore, P. and Robinson, A. (2016). 'The Quantified Self: What Counts in the Neoliberal Workplace'. *New Media & Society*, 18(11): 2774–92.
13. Whitson, J. R. (2013). 'Gaming the Quantified Self'. *Surveillance & Society*, 11(1/2): 163.
14. See, for instance. Elias, A. S. and Gill, R. (2017). 'Beauty Surveillance: The Digital Self-monitoring Cultures of Neoliberalism'. *European Journal of Cultural Studies*, 21(1): 59–77; Sian, K. P. (2015). 'Spies, Surveillance and Stakeouts: Monitoring Muslim Moves in British State Schools'. *Race Ethnicity and Education*, 18(2): 183–201.
15. See Andrejevic, M. (2007). *iSpy: Surveillance and Power in the Interactive Era*. University Press of Kansas: 182.
16. Moore, P. V. (2017). *The Quantified Self in Precarity: Work, Technology and What Counts*. Routledge.

17. Till, C. (2014). 'Exercise As Labour: Quantified Self and the Transformation of Exercise into Labour'. *Societies*, 4(3): 446–62.
18. Ajana, B. (2017). 'Digital Health and the Biopolitics of the Quantified Self'. *Digital Health*, 3.
19. Quoted in Attili, S. (2014). 'Data is the New Natural Resource'. *Fortune*, 16 March.
20. Anon, 'Why Big Data is the New Natural Resource'. *Washington Post.*
21. Crooks, E. (2018). 'Drillers Turn to Big Data in the Hunt for More, Cheaper Oil'. *Financial Times*, 12 February.
22. Bernard, M. (2017). 'Why Space Data is the New Big Data'. *Fortune*, 19 October.
23. Dodson, S. (2014). 'Big Data, Big Hype?'. *Wired.*
24. Lugmayr, A., Stockleben, B., Scheib, C. and Mailaparampil, M. A. (2017). 'Cognitive Big Data: Survey and Review on Big Data Research and Its Implications. What is Really "New" in Big Data?'. *Journal of Knowledge Management*, 21(1): 197–212.
25. Anon (2017). 'Data is Giving Rise to a New Economy'. *The Economist*, 6 May.
26. Schlosser, A. (2018). 'You May Have Heard Data is the New Oil: It's Not'. *World Economic Forum*, 10 January.
27. Tarnoff, B. (2018). 'Big Data for the People: It's Time to Take it Back from Our Tech Overlords'. *The Guardian*, 14 March.
28. Higgs, E. (2001). 'The Rise of the Information State: The Development of Central State Surveillance of the Citizen in England, 1500–2000'. *Journal of Historical Sociology*, 14(2): 175–97.
29. Marglin, S. A. (1974). 'What Do Bosses Do? the Origins and Functions of Hierarchy in Capitalist Production'. *Review of Radical Political Economics*, 6(2): 60–112.
30. Jensen, M. C. (1993). 'The Modern Industrial Revolution, Exit, and the Failure of Internal Control Systems'. *The Journal of Finance*, 48(3): 831–80. Also see Burawoy, M. (1983). 'Between the Labor Process and the State: The Changing Face of Factory Regimes Under Advanced Capitalism'. *American Sociological Review*, 48(5): 587–605; Burawoy, M. (1984). 'Karl Marx and the Satanic Mills: Factory Politics Under Early Capitalism in England, the United States, and Russia'. *American Journal of Sociology*, 90(2): 247–82.
31. Lyon, D. (1994). *The Electronic Eye: The Rise of Surveillance Society.* University of Minnesota Press.

32. Gandy Jr, O. H. (1989). 'The Surveillance Society: Information Technology and Bureaucratic Social Control'. *Journal of Communication*, 39(3): 61–76.
33. See Brayne, S. (2017). 'Big Data Surveillance: The Case of Policing'. *American Sociological Review*, 82(5): 977–1008.
34. West, S. M. (2017). 'Data Capitalism: Redefining the Logics of Surveillance and Privacy'. *Business & Society*. Online first publication.
35. John Walker, S. (2014). 'Big Data: A Revolution that Will Transform How we Live, Work, and Think'. *International Journal of Advertising*, 33(1): 181–3.
36. Boyd, D. and Crawford, K. (2012). 'Critical Questions for Big Data: Provocations for a Cultural, Technological, and Scholarly Phenomenon'. *Information, Communication & Society*, 15(5): 662–79.
37. Tene, O. and Polonetsky, J. (2013). 'Big Data for All: Privacy and User Control in the Age of Analytics'. *Northwestern Journal of Technology and Intellectual Property* 11(5).
38. Langley, P. and Leyshon, A. (2017). 'Platform Capitalism: The Intermediation and Capitalisation of Digital Economic Circulation'. *Finance and Society*, 3(1): 11–31.
39. See Clarke, R. (1988). 'Information Technology and Dataveillance'. *Communications of the ACM*, 31(5): 498–512; Van Dijck, J. (2014). 'Datafication, Dataism and Dataveillance: Big Data Between Scientific Paradigm and Ideology'. *Surveillance & Society*, 12(2): 197; Degli Esposti, S. (2014). 'When Big Data Meets Dataveillance: The Hidden Side of Analytics'. *Surveillance & Society*, 12(2): 209.
40. Wood, D. M. and Ball, K. (2013). 'Brandscapes of Control? Surveillance, Marketing and the Co-construction of Subjectivity and Space in Neo-liberal Capitalism'. *Marketing Theory*, 13(1): 47–67.
41. Dyer-Witheford, N. (2015). *Cyber-proletariat: Global Labour in the Digital Vortex*. Pluto: 5.
42. O'Neil, C. (2016). *Weapons of Math Destruction*. Penguin.
43. Turner, B. S., Abercrombie, N. and Hill, S. (2014). *Sovereign Individuals of Capitalism*. Routledge.
44. Pecora, V. P. (2002). 'The Culture of Surveillance'. *Qualitative Sociology*, 25(3): 345–58.

45. Christian, B. and Griffiths, T. (2016). *Algorithms to Live By: The Computer Science of Human Decisions*. Macmillan.
46. Manovich, L. (2011). 'Trending: The Promises and the Challenges of Big Social Data'. *Debates in the Digital Humanities*, 2: 460–75.
47. See Taplin, J. (2017). *Move Fast and Break Things: How Facebook, Google, and Amazon Have Cornered Culture and What It Means for All of Us*. Macmillan.
48. Ford, M. (2015). *The Rise of the Robots: Technology and the Threat of Mass Unemployment*. Oneworld Publications.
49. Bartlett, J. (2018). *The People vs Tech: How the Internet is Killing Democracy (and How We Save It)*. Penguin.
50. Townsend, A. M. (2013). *Smart Cities: Big Data, Civic Hackers, and the Quest for a New Utopia*. W.W. Norton and Company.
51. Söderström, O., Paasche, T. and Klauser, F. (2014). 'Smart Cities As Corporate Storytelling'. *City*, 18(3): 307–20.
52. Datta, A. (2015). 'New Urban Utopias of Postcolonial India: "Entrepreneurial Urbanization" in Dholera Smart City, Gujarat'. *Dialogues in Human Geography*, 5(1): 3–22.
53. Troullinou, P. (2017). 'Exploring the Subjective Experience of Everyday Surveillance: The Case of Smartphone Devices As Means of Facilitating "Seductive" Surveillance'. Doctoral dissertation, Open University.
54. McAfee, A. and Brynjolfsson, E. (2017). *Machine, Platform, Crowd: Harnessing Our Digital Future*. W.W. Norton and Company.
55. Vanolo, A. (2014). 'Smartmentality: The Smart City As Disciplinary Strategy'. *Urban Studies*, 51(5): 883.
56. See Harper, D., Tucker, I. and Ellis, D. (2013). 'Surveillance and Subjectivity: Everyday Experiences of Surveillance Practices'. In Ball, K. and Snider, L. (eds): *The Surveillance-Industrial Complex: A Political Economy of Surveillance*. Routledge.
57. Zurawski, N. (2011). 'Local Practice and Global Data: Loyalty Cards, Social Practices, and Consumer Surveillance'. *The Sociological Quarterly*, 52(4): 509–27: 509.
58. Kealy, H. (2014). 'The Apps Designed to Keep Your Teen Under Control'. *The Telegraph*, 21 August.
59. Ibid.
60. Williams, R. (2015). 'Spyware and Smartphones: How Abusive Men Track their Partners'. *The Guardian*, 25 January.
61. See Hayes B. (2012) 'The Surveillance-industrial Complex'. In Ball K., Haggerty K. and Lyon D. (eds): *Routledge Handbook of*

Surveillance Studies. Routledge: 167; Fuchs, C. (2016). Information Ethics in the Age of Digital Labour and the Surveillance-industrial Complex'. In Kelly, M. and Bielby, J. (eds): *Information Cultures in the Digital Age*. Springer.

62. Marx, K. (1976). *Capital: A Critique of Political Economy*, Volume I (1867): Harmondsworth, London: 342, 367.
63. Thrift, N. (2005). *Knowing Capitalism*. Sage.
64. Ibid., 5.
65. Stanley, L. (2008). 'It Has Always Known, and We Have Always Been "Other": Knowing Capitalism and the "Coming Crisis" of Sociology Confront the Concentration System and Mass-observation'. *The Sociological Review*, 56(4): 535–51.
66. Beer, D. (2009). 'Power Through the Algorithm? Participatory Web Cultures and the Technological Unconscious'. *New Media & Society*, 11(6): 985–1002.
67. Kitchin, R. (2014). *The Data Revolution: Big Data, Open Data, Data Infrastructures and Their Consequences*. Sage.
68. Foster, J. B. and McChesney, R. W. (2014). 'Surveillance Capitalism: Monopoly-finance Capital, the Military-industrial Complex, and the Digital Age'. *Monthly Review*, 66(3): 18.
69. Fuchs, C. and Trottier, D. (2015). 'Towards a Theoretical Model of Social Media Surveillance in Contemporary Society'. *Communications: The European Journal of Communication Research*, 40(1): 113–35.
70. Wilson, D. and McCulloch, J. (2015). *Pre-crime: Pre-emption, Precaution and the Future*. Routledge.
71. See Ball, K. and Webster, F. (2003). *The Intensification of Surveillance: Crime, Terrorism and Warfare in the Information Era*. Pluto.
72. Weizman, E. (2002). Introduction to The Politics of Verticality'. *Open Democracy*. Also Bracken-Roche, C. (2016). 'Domestic Drones: The Politics of Verticality and the Surveillance Industrial Complex'. *Geographica Helvetica*, 71(3): 167–72.
73. Wood, D. M. (2013). 'What is Global Surveillance? Towards a Relational Political Economy of the Global Surveillant Assemblage'. *Geoforum*, 49: 317–26.
74. Zimmer, M. (2008). 'The Gaze of the Perfect Search Engine: Google As an Infrastructure of Dataveillance'. In Spink, A. and Zimmer, M. (eds): *Web Search*. Springer.

75. See Coté, M. (2014). 'Data Motility: The Materiality of Big Social Data'. *Cultural Studies Review*, 20(1): 121; Pasquinelli, M. (2009). 'Google's Pagerank Algorithm: A Diagram of Cognitive Capitalism and the Rentier of the Common Intellect'. *Deep Search*, 3: 152–62.
76. Mohan, S. (2016). '"Big Data is Like Sex": Here's a Peek into What Firms Desire and Actually Achieve'. *Financial Express*, 6 June.
77. See Spitzer, S. (1979). 'The Rationalization of Crime Control in Capitalist Society'. *Contemporary Crises*, 3(2): 187–206; Eyles, J. (1985). 'From Equalisation to Rationalisation: Public Health Care Provision in New South Wales'. *Geographical Research*, 23(2): 243–68; Nyland, C. (1985). 'Worktime and the Rationalisation of the Capitalist Production Process'. Doctoral dissertation.
78. Lash, S. and Urry, J. (1987). *The End of Organized Capitalism*. University of Wisconsin Press.
79. See Steele, B. J. (2005). 'Ontological Security and the Power of Self-identity: British Neutrality and the American Civil War'. *Review of International Studies*, 31(3): 519–40; Mitzen, J. (2006). 'Ontological Security in World Politics: State Identity and the Security Dilemma'. *European Journal of International Relations*, 12(3): 341–70; Brown, W. S. (2000). 'Ontological Security, Existential Anxiety and Workplace Privacy'. *Journal of Business Ethics*, 23(1): 61–65.
80. Giddens, A. (1991). *Modernity and Self-identity: Self and Society in the Late Modern Age*. Polity Press.
81. See for instance Block, F. and Somers, M. R. (2014). *The Power of Market Fundamentalism*. Harvard University Press; Stiglitz, D. J. (2009). 'Moving Beyond Market Fundamentalism to a More Balanced Economy'. *Annals Of Public and Cooperative Economics*, 80(3): 345–60; Kozul-Wright, R. and Rayment, P. B. W. (2007). *The Resistible Rise of Market Fundamentalism: Rethinking Development Policy in an Unbalanced World*. Third World Network.
82. See Welch, M. (2006). *Scapegoats of September 11th: Hate Crimes and State Crimes in the War on Terror*. Rutgers University Press; Gurtov, M. (2006). *Superpower on Crusade: The Bush Doctrine in US Foreign Policy*. Lynne Rienner Publishers: 35.
83. See Ritzer, G. (2015). 'Prosumer Capitalism'. *The Sociological Quarterly*, 56(3): 413–45; Zwick, D. (2015). 'Defending the Right Lines of Division: Ritzer's Prosumer Capitalism in the Age of

Commercial Customer Surveillance and Big Data'. *The Sociological Quarterly*, 56(3): 484–98.

84. Comor, E. (2010). 'Digital Prosumption and Alienation'. *Ephemera*, 10(3): 439.
85. Pongratz, H. J. and Voß, G. G. (2003). 'From Employee to "Entreployee": Towards a "Self-entrepreneurial" Work Force?'. *Concepts and Transformation*, 8(3): 239–54; Also see Marr, B. (2017). *Data Strategy: How to Profit from a World of Big Data, Analytics and the Internet of Things*. Kogan Page.
86. Lyon, D. (2005). 'Surveillance As Social Sorting: Computer Codes and Mobile Bodies'. In *Surveillance as Social Sorting*. Routledge.
87. See for instance Bain, P. and Taylor, P. (2000). 'Entrapped by the "Electronic Panopticon"? Worker Resistance in the Call Centre'. *New Technology, Work and Employment*, 15(1): 2–18.
88. Foucault, M. (1995 [1977]). *Discipline and Punish: The Birth of the Prison*. Translated by Alan Sheridan. Random House: 201.
89. Ibid., 220.
90. O'Neill, J. (1986). 'The Disciplinary Society: From Weber to Foucault'. *British Journal of Sociology*, 37(1): 42–60. Also see 'The Post-panoptic Society? Reassessing Foucault in Surveillance Studies'. *Social Identities*, 16(5): 621–33.
91. Haggerty, K. D. and Ericson, R. V. (2000). 'The Surveillant Assemblage'. *The British Journal of Sociology*, 51(4): 605–22.
92. Nygren, K. G. and Gidlund, K. L. (2015). 'The Pastoral Power of Technology: Rethinking Alienation in Digital Culture'. *TripleC*, 10(2): 509–17.
93. Fernie, S. and Metcalf, D. (1998). *(Not) Hanging on the Telephone: Payment Systems in the New Sweatshops*. Centre for Economic Performance, London School of Economics and Political Science: 9.
94. Žižek, S. (1993). *Tarrying with the Negative*. Duke University Press: 201.
95. Mathiesen, T. (1997). 'The Viewer Society: Michel Foucault's Panopticon Revisited'. *Theoretical Criminology*, 1(2): 217.

Chapter 3

1. Bauman, Z. (2000). *Liquid Modernity*. Polity: 6.
2. Ibid., 6.

3. See Hill Collins, P. and Bilge, S. (2016). *Intersectionality*. Wiley; Crenshaw, K. (1991). 'Mapping the Margins: Intersectionality, Identity Politics, and Violence against Women of Color'. *Stanford Law Review*, 43(6): 1241–99; Hancock, A. (2007). 'When Multiplication Doesn't Equal Quick Addition: Examining Intersectionality as a Research Paradigm'. *Perspectives on Politics*, 5(1): 63–79.
4. Eisner, S. (2013). *Bi: Notes for a Bisexual Revolution*. Seal Press.
5. Castells, M. (1997). *The Power of Identity*. Blackwell: 2.
6. Ibid., 2.
7. Gergen, K. (2002). *The Saturated Self*. Basic Books.
8. Ibid., 3.
9. Cover, R. and Prosser, R. (2013). 'Memorial Accounts: Queer Young Men, Identity and Contemporary Coming Out Narratives Online'. *Australian Feminist Studies*, 28(75): 81–94.
10. Foucault, M. (1988). 'Technologies of the Self'. In Martin, L., Gutman, H. and Hutton, P. (eds), *Technologies of the Self*. University of Massachusetts Press: 16.
11. Ibid., 30.
12. Mennicken, A. and Miller, P. (2014). 'Michel Foucault and the Administering of Lives'. In Adler, P., Du Gay, P., Morgan, G. and Reed, M. (eds), *The Oxford Handbook on Sociology, Social Theory and Organisation Studies*. Oxford University Press: 13.
13. Cooper, C. (2015). 'Entrepreneurs of the Self: The Development of Management Control Since 1976'. *Accounting, Organizations and Society*, 47: 14–24.
14. Wallace, E., Buil, I. and de Chernatony, L. (2012). 'Facebook "Friendship" and Brand Advocacy'. *Journal of Brand Management*, 20(2): 128.
15. Wang, X., Yu, C. and Wei, Y. (2012). 'Social Media Peer Communication and Impacts on Purchase Intentions: A Consumer Socialization Framework'. *Journal of Interactive Marketing*, 26(4): 198–208.
16. Lupton, D. and Seymour, W. (2000). 'Technology, Selfhood and Physical Disability'. *Social Science & Medicine*, 50(12): 1851–62.
17. Garza, A., Tometi, O. and Cullors, P. (2014). 'A Herstory of the #Black Lives Matter Movement'. In Hobson, J. (ed.), *Are All the Women Still White: Rethinking Race, Expanding Feminisms*. State University of New York Press: 23.

18. Bennett, W. (2012). 'The Personalization of Politics: Political Identity, Social Media, and Changing Patterns of Participation. *The ANNALS of the American Academy of Political and Social Science*, 644(1): 20.
19. Laclau, E. (1996). 'The Death and Resurrection of the Theory of Ideology'. *Journal of Political Ideologies*, 1(3): 201–20.
20. Barber, B. (2000). *Jihad vs. McWorld*. MTM.
21. Tuckle, S. (1995). *Life on the Screen*. Simon and Schuster.
22. Silva, P., Holden, K. and Nii, A. (2014). 'Smartphones, Smart Seniors, But Not-So-Smart Apps: A Heuristic Evaluation of Fitness Apps'. In Schmorrow, D. and Fidopiastis, C. (eds), *Foundations of Augmented Cognition: Advancing Human Performance and Decision-making through Adaptive Systems*. Springer.
23. See Brown, A. (2015). 'Identities and Identity Work in Organizations'. *International Journal of Management Reviews*, 17(1): 20–40; Casey, C. (1995). *Work, Self and Society: After Industrialisation*. Taylor and Francis; Fleming, P. and Spicer, A. (2003). 'Working at a Cynical Distance: Implications for Power, Subjectivity and Resistance'. *Organization*, 10(1): 157–79.
24. Sveningsson, S. and Alvesson, M. (2003). 'Managing Managerial Identities: Organizational Fragmentation, Discourse and Identity Struggle'. *Human Relations*, 56(10): 1165.
25. See Beck, U. (1997). *The Reinvention of Politics*. Polity. Boxenbaum, E. and Rouleau, L. (2011). 'New Knowledge Products as Bricolage: Metaphors and Scripts in Organizational Theory'. *Academy of Management Review*, 36(2): 272–96.
26. Ma, M. and Agarwal, R. (2007). 'Through a Glass Darkly: Information Technology Design, Identity Verification, and Knowledge Contribution in Online Communities'. *Information Systems Research*, 18(1): 42–67.
27. Hearn, A. (2017). 'Verified: Self-presentation, Identity Management, and Selfhood in the Age of Big Data'. *Popular Communication*, 15(2): 62–77.
28. Cullen, J. (2009). 'How to Sell Your Soul and Still Get into Heaven: Steven Covey's Epiphany-inducing Technology of Effective Selfhood'. *Human Relations*, 62(8): 1231–54.
29. Paruchuri,, V. and Chellappan, S. (2013). 'Context Aware Identity Management Using Smart Phones'. In *Broadband and Wireless*

Computing, Communication and Applications (BWCCA): 2013 Eighth International Conference On.
30. See Sullivan, J. (2013). 'How Google Is Using People Analytics to Completely Reinvent HR'. *Talent Management and HR*; Bryant, A. (2011). 'Google's Quest to Build a Better Boss'. *New York Times.*
31. Fecheyr-Lippens, B., Schaninger, B. and Tanner, K. (2015). 'Power to the New People Analytics'. *McKinsey Quarterly*: 1.
32. Isson, J. and Harriott, J. (2016). *People Analytics in the Era of Big Data: Changing the Way You Attract, Develop, and Retain Talent.* John Wiley and Sons: xv.
33. Burchielli, R., Bartram, T. and Thanacoody, R. (2008). 'Work-Family Balance or Greedy Organizations?'. *Relations Industrielles*, 63(1): 108.
34. Hardey, M. (2002). '"The Story of My Illness": Personal Accounts of Illness on the Internet'. *Health*, 6(1): 31–46.
35. Busold, C., Heuser, S., Rios, J., Sadeghi, A. and Asoken, N. (2015). 'Smart and Secure Cross-Device Apps for the Internet of Advanced Things'. In Böhme, R. and Okamoto, T. (eds), *Financial Cryptography and Data Security*. Springer: 1.
36. Grace, R. (2015). 'These Eight Work-Life Balance Apps Will Transform Your Business and Life'. *Huffington Post.*
37. See Gotsi, M., Andriopoulos, C., Lewis, M. and Ingram, A. (2010). 'Managing Creatives: Paradoxical Approaches to Identity Regulation'. *Human Relations*, 63(6): 781–805; Wasserman, V. and Frenkel, M. (2011). 'Organizational Aesthetics: Caught Between Identity Regulation and Culture Jamming'. *Organization Science*, 22(2): 503–21.
38. Alvesson, M. and Willmott, H. (2002). 'Identity Regulation as Organizational Control: Producing the Appropriate Individual'. *Journal of Management Studies*, 39(5): 620.
39. Goffman, E. (1959). *The Presentation of Self in Everyday Life.* Anchor Books.
40. Zhao, S., Grasmuck, S. and Martin, J. (2008). 'Identity Construction on Facebook: Digital Empowerment in Anchored Relationships'. *Computers in Human Behavior*, 24(5): 1816.
41. Lupton, D. (2012). 'M-health and Health Promotion: The Digital Cyborg and Surveillance Society'. *Social Theory & Health*, 10(3): 229–44.

42. Gardner, H. and Davis, K. (2013). *The App Generation*. Yale University Press: 66.
43. Fleming, P. and Sturdy, A. (2009). 'Just Be Yourself!'. *Employee Relations*, 31(6): 569–83.
44. Bloom, P. (2013). 'Fight for Your Alienation: The Fantasy of Employability and the Ironic Struggle for Self-exploitation'. *Ephemera*, 13(4): 785–807.
45. Foucault, M. and Sheridan, A. (1977). *Discipline and Punish*. Vintage: 215.
46. Ibid., 138.
47. Krahnke, K., Giacalone, R. and Jurkiewicz, C. (2003). 'Point-Counterpoint: Measuring Workplace Spirituality'. *Journal of Organizational Change Management*, 16(4): 397.
48. Willmott, H. (1993). 'Strength is Ignorance; Slavery is Freedom: Managing Culture in Modern Organizations'. *Journal of Management Studies*, 30(4): 515–52.
49. MacLullich, K. (2003). 'The Emperor's "New" Clothes? New Audit Regimes: Insights from Foucault's Technologies of the Self'. *Critical Perspectives on Accounting*, 14(8): 791.
50. Gill, R. (2017). '"Life is a Pitch": Managing the Self in New Media Work'. In Deuze, M. (ed.), *Managing Media Work*. Sage.
51. Lacan, J. (2000). *Ecrites*. W.W. Norton and Company.
52. Žižek, S. (1993). *Tarrying with the Negative*. Duke University Press Books: 201.
53. Stavrakakis, Y. (2012). *Lacan and the Political*. Taylor and Francis: 29.
54. Van Zoonen, L. and Turner, G. (2013). 'Taboos and Desires of the UK Public for Identity Management in the Future: Findings from Two Survey Games'. In *Proceedings of the 2013 ACM Workshop on Digital Identity Management*: 37.
55. Costas, J. and Fleming, P. (2009). 'Beyond Dis-identification: A Discursive Approach to Self-alienation in Contemporary Organizations'. *Human Relations*, 62(3): 353–78; Also see Leidner, R. (2006). *Fast Food, Fast Talk*. University of California Press; Sennett, R. (1998). *The Corrosion of Character: The Personal Consequences of Work in The New Capitalism*. W.W. Norton.
56. Mumby, D. (2005). 'Theorizing Resistance in Organization Studies: A Dialectical Approach'. *Management Communication Quarterly*, 19(1): 19–44.

57. Zimmerman, M. (2000). 'The End of Authentic Selfhood in the Postmodern Age'. In Wrathall, M. and Malpas, J. (eds), *Heidegger, Authenticity, and Modernity: Essays in Honour of Dreyfus*, MIT Press: 123.
58. Dean, M. (1994). '"A Social Structure of Many Souls": Moral Regulation, Government, and Self-Formation'. *Canadian Journal of Sociology*, 19(2): 145; Rimke, H. (2000). 'Governing Citizens Through Self-Help Literature'. *Cultural Studies*, 14(1): 61–78; Rose, N. (1989). *Inventing Our Selves*. Cambridge University Press.
59. Cremin, C. (2009). 'Never Employable Enough: The (Im) possibility of Satisfying the Boss's Desire'. *Organization*, 17(2): 131–49.
60. Field, F. (1997). 'Re-inventing Welfare: A Response to Lawrence Mead'. In Mead, L. (ed.), *From Welfare To work: Lessons from America*. Institute for Economic Affairs, Health, Welfare and Work: 62.
61. Kanter, R. (1994). 'Employability and Job Security in the 21st Century'. *Demos*, 1.
62. Bauman, Z. (2013). *Liquid Modernity*. John Wiley and Sons.
63. McLaren, D. and Agyeman, J. (2015). *Sharing Cities*. MIT Press: 3.
64. Cederstrom, C. and Spicer, A. (2015). *The Wellness Syndrome*. Polity.

Chapter 4

1. Tardanico, S. (2012). 'Is Social Media Sabotaging Real Communication?'. *Forbes*, 30 April.
2. Anderson, B. (2006). *Imagined Communities: Reflections on the Origin and Spread of Nationalism*. Verso.
3. Hampton, K. (2012). 'Social Media as Community'. *New York Times*, 18 June.
4. Ritzer, G. (2004). 'The McDonaldization of Society'. *Newbury Park*: 170.
5. See Goggin, G. (2012). 'The Eccentric Career of Mobile Television'. *International Journal of Digital Television*, 3(2): 119–40; Wilken, R. and Goggin, G. (2014). *Mobile Technology and Place*. Routledge.

6. See Carroll, N., Richardson, I. and Whelan, E. (2012). 'Service Science'. *International Journal of Actor-network Theory and Technological Innovation*, 4(3): 51–69; Latour, B. (1987). *Science in Action*. Harvard University Press.
7. Woolgar, S. (2005). 'Mobile Back to Front: Uncertainty and Danger in the Theory-technology Relation'. In Ling, R. and Pederson, P. (eds), *Mobile Communications: Renegotiation of the Social Sphere*. Springer.
8. Newman, A. and Zysman, J. (2006). 'Frameworks for Understanding the Political Economy of the Digital Era'. In Newman, A. and Zysman, J. (eds), *How Revolutionary Was the Digital Revolution? National Responses, Market Transitions and Global Technology*. Stanford University Press: 4.
9. Jameson, F. (1985). 'Postmodernism and Consumer Society'. *Postmodern Culture*: 111.
10. Wu, M. and Pearce, P. (2013). 'Appraising Netnography: Towards Insights About New Markets in the Digital Tourist Era'. *Current Issues in Tourism*, 17(5): 463–74.
11. Jenkins, H., Ford, S. and Green, J. (2013). *Spreadable Media*. New York University Press: 1.
12. Lessig, L. (1996). 'The Zones of Cyberspace'. *Stanford Law Review*, 48(5): 1403.
13. Lefebvre, H. (1991). *The Production of Social Space*. Basil Blackwell: 26.
14. Shields, R. (1999). *Lefebvre, Love and Struggle*. Routledge.
15. Lefebvre, H., Elden, S. and Moore, G. (2004). *Rhythmanalysis*. Bloomsbury Academic.
16. Cohen, J. (2007). 'Cyberspace as/and Space'. *Columbia Law Review*, 107(1): 210.
17. Samuels, R. (2008). 'Auto-modernity After Postmodernism: Autonomy and Automation in Culture, Technology, and Education'. In McPherson, T. (ed.), *Digital Youth, Innovation, and the Unexpected.* MIT Press, 219–40.
18. Munro, R. (1997).'The Consumption View of Self: Extension, Exchange and Identity'. *The Sociological Review*, 44(1 suppl.): 248–73.
19. Thompson, C. (2005). 'Meet the Life Hackers'. *New York Times*.
20. Trapani, G. (2008). *Upgrade Your Life*. John Wiley and Sons.
21. See Lockie, S. and Higgins, V. (2007). 'Roll-out Neoliberalism and Hybrid Practices of Regulation in Australian Agri-environmental

Governance'. *Journal of Rural Studies*, 23(1): 1–11; Peck, J. and Tickell, A. (2002). 'Neoliberalizing Space'. *Antipode*, 34(3): 380–404.

22. Reynolds, L. and Szerszynsk, B. (2012). 'Neoliberalism and Technology: Perpetual Innovation or Perpetual Crisis?'. In Pellizzoni, L. and Ylönen, M. (eds), *Neoliberalism and Technoscience: Critical Assessments*. Ashgate: 29.
23. Goggin, G. (2012). *Cell Phone Culture*. Taylor and Francis: 206–7.
24. Odendall, N. (2016). 'Smart City: Neoliberal Discourse or Urban Development Tool?'. In Grugel, J. and Hammett, D. (eds), *The Palgrave Handbook of International Development*. Palgrave Macmillan: 615.
25. Bourdieu, P. (1984). *Distinction: A Social Critique of the Judgement of Taste*. Routledge: 170.
26. Crary, J. (2013). *24/7: Late Capitalism and the Ends of Sleep*. Verso: 3–4.
27. Jarvenpaa, S. and Lang, K. (2005). 'Managing the Paradoxes of Mobile Technology'. *Information Systems Management*, 22(4): 7–23: 11.
28. Ibid., 7–23.
29. Binkley, S. (2009). 'The Work of Neoliberal Governmentality: Temporality and Ethical Substance in the Tale of Two Dads'. *Foucault Studies*, 6: 60.
30. See Arthur, M. and Rousseau, D. (2001). *The Boundaryless Career*. Oxford University Press; Sullivan, S. and Arthur, M. (2006). 'The Evolution of the Boundaryless Career Concept: Examining Physical and Psychological Mobility'. *Journal of Vocational Behavior*, 69(1): 19–29.
31. Marx, K. (1976). *Capital: A Critique of Political Economy*, Volume 1 (1867). Penguin: 324; Also see Bloom, P. (2014). 'We Are All Monsters Now!'. *Equality, Diversity and Inclusion: An International Journal*, 33(7): 662–80; Neocleous, M. (2003). 'The Political Economy of the Dead Marx's Vampires'. *History of Political Thought*, 24(4): 668–84.
32. Marx, K. (1976). *Capital: A Critique of Political Economy*, Volume 1 (1867). Penguin: 945–6.
33. Hobsbawm, E. (2010). *Age of Empire*. Weidenfield and Nicolson.
34. Hollands, R. (2014). 'Critical Interventions into the Corporate Smart City'. *Cambridge Journal of Regions, Economy and Society*, 8(1): 61.

35. Ortega, A. (2012). 'Desakota and Beyond: Neoliberal Production of Suburban Space in Manila's Fringe'. *Urban Geography*, 33(8): 1118–43.
36. Foucault, M. (1984). 'Of Other Spaces: Utopias and Heterotopias'. *Architecture/Mouvement/Continuité*, 5: 46–9.
37. Ong, A. (2007). 'Neoliberalism As a Mobile Technology'. *Transactions of the Institute of British Geographers*, 32(1): 3.
38. Lombardi, P. and Vanolo, A. (2015). 'Smart City As a Mobile Technology: Critical Perspectives on Urban Development Policies'. In Rodrìguez-Bolívar, M. (ed.), *Transforming City Governments for Successful Smart Cities*. Springer: 147.
39. Boltanski, L. and Chiapello, E. (2007). 'The New Spirit of Capitalism'. *Capital & Class*, 31(2): 198–201.
40. Žižek, S. (2004). 'What can Psychoanalysis Tell Us About Cyberspace?'. *The Psychoanalytic Review*, 91(6): 801–30.
41. Przybylski, A., Murayama, K., DeHaan, C. and Gladwell, V. (2013). 'Motivational, Emotional, and Behavioral Correlates of Fear of Missing Out'. *Computers in Human Behavior*, 29(4): 1841–8.
42. Mitchell, T. (1999). 'Dreamland: The Neoliberalism of Your Desires'. *Middle East Report*, (210): 28.
43. Sparke, M. (2006). 'A Neoliberal Nexus: Economy, Security and the Biopolitics of Citizenship on the Border'. *Political Geography*, 25(2): 151–80.
44. Bauman, Z. (2000). *Liquid Modernity*. Polity: 7.

Chapter 5

1. Wucker, M. (2018) 'How to Have a Good Fourth Industrial Revolution'. *World Economic Forum*, 21 June.
2. Spicer, A. and Cederstrom, C. (2015). 'The Dark Underbelly of the Davos "Well-being" Agenda'. *Washington Post*. This insight was actually building on the original ideas of Chrisopher Lasch from the 1970s in Lasch, C. (1976). 'The Narcissist Society'. *New York Review of Books*, 30 September.
3. Klein, N. (2016). 'It Was the Democrats' Embrace of Neoliberalism that Won it for Trump'. *The Guardian*, 9 November.
4. See Duara, P. (2001). 'The Discourse of Civilization and Pan-Asianism'. *Journal of World History*, 12(1): 99–130; Ashcroft, B.

(2013). 'Post-colonial Transformation'. Routledge; Dussel, E. D., Krauel, J., and Tuma, V. C. (2000). 'Europe, Modernity, and Eurocentrism'. *Nepantla: Views from South*, 1(3): 465–78; Cooper, F. and Stoler, A. L. (eds) (1997). *Tensions of Empire: Colonial Cultures in a Bourgeois World*. University of California Press; Linklater, A. (2005). 'A European Civilising Process'. In Hill, C. and Smith, M. (eds), *International Relations and the European Union*. Oxford University Press.

5. Quote in Bloom, P. (2016). *Beyond Power and Resistance: Politics at the Radical Limits*. Rowman and Littlefield International.
6. See for instance Thompson, F. M. L. (1981). 'Social Control in Victorian Britain'. *The Economic History Review*, 34(2): 189–208; Jones, G. S. (2014). *Outcast London: A Study in the Relationship Between Classes in Victorian Society*. Verso; Searle, G. R. (1998). *Morality and the Market in Victorian Britain*. Oxford University Press; Ignatieff, M. (1978). A Just Measure of Pain: The Penitentiary in the Industrial Revolution, 1750–1850. Macmillan.
7. Weber, M. (2002). *The Protestant Ethic and the Spirit of Capitalism*. Wilder Publications; also see Furnham, A., Bond, M., Heaven, P., Hilton, D., Lobel, T., Masters, J., et al. (1993). 'A Comparison of Protestant Work Ethic Beliefs in Thirteen Nations'. *The Journal of Social Psychology*, 133(2): 185–97. Giorgi, L. and Marsh, C. (1990). 'The Protestant Work Ethic as a Cultural Phenomenon'. *European Journal of Social Psychology*, 20(6): 499–517.
8. See Bowler, K. (2018). *Blessed: A History of the American Prosperity Gospel*. Oxford University Press; Koch, B. A. (2010). 'The Prosperity Gospel and Economic Prosperity: Race, Class, Giving, and Voting'. PhD dissertation, Indiana University; Lioy, D. (2007). 'The Heart of the Prosperity Gospel: Self or the Savior?'. *Conspectus: The Journal of the South African Theological Seminary*, 4(9): 41–64.
9. See Gaillard, A. and Garaudi, R. (1974). 'Christianity and Marxism'. No. CERN-AUDIO-1974-004; MacIntyre, A. (1984). *Marxism and Christianity*. University of Notre Dame Press; Löwy, M. (1993). 'Marxism and Christianity in Latin America'. *Latin American Perspectives*, 20(4): 28–42.
10. See Perry, L. (1973). Radical Abolitionism: Anarchy and the Government of God in Antislavery Thought. University of Tennessee Press; Stewart, J. B. and Foner, E. (1996). Holy Warriors: The Abolitionists and American Slavery. Macmillan; Emerson, M.

O. and Smith, C. (2001). Divided by Faith: Evangelical Religion and the Problem of Race in America. Oxford University Press.

11. See Houck, D. W. and Dixon, D. E. (eds) (2006). *Rhetoric, Religion and the Civil Rights Movement, 1954–1965*, Volume 2. Baylor University Press; Smith, C. (ed.) (2014). *Disruptive Religion: The Force of Faith in Social Movement Activism*. Routledge; Hunt, L. L. and Hunt, J. G. (1977). 'Black Religion As Both Opiate and Inspiration of Civil Rights Militance: Putting Marx's Data to the Test'. *Social Forces*, 56(1): 1–14.
12. See Berryman, P. (1987). *Liberation Theology*. McGraw-Hill.; Smith, C. (1991). *The Emergence of Liberation Theology: Radical Religion and Social Movement Theory*. University of Chicago Press; Berryman, P. (1987). *Liberation Theology: Essential Facts About the Revolutionary Movement in Latin America – and Beyond*. Temple University Press; Boff, L. (2012). *Church: Charism and Power: Liberation Theology and the Institutional Church*. Wipf and Stock Publishers.
13. Mihevc, J. (1995). *The Market Tells Them So: The World Bank and Economic Fundamentalism in Africa*. Zed Books; also Hicks, A. (2006). 'Free-market and Religious Fundamentalists Versus Poor Relief'. *American Sociological Review*, 71(3): 503–10.
14. Boldeman, L. (2013). *The Cult of the Market: Economic Fundamentalism and its Discontents*. ANU Press: 316.
15. See Lewis, M. K. and Algaoud, L. M. (2001). *Islamic Banking*. Edward Elgar; Maurer, B. (2005). *Mutual Life, Limited: Islamic Banking, Alternative Currencies, Lateral Reason*. Princeton University Press.
16. Stiglitz, D. (2009). 'Moving Beyond Market Fundamentalism to a More Balanced Economy'. *Annals of Public and Cooperative Economics*, 80(3): 345–60.
17. Boltanski, L. and Chiapello, E. (2005). 'The New Spirit of Capitalism'. *International Journal of Politics, Culture, and Society*, 18(3–4): 161–88.
18. See Cooper, M. E. (2011). *Life as Surplus: Biotechnology and Capitalism in the Neoliberal Era*. University of Washington Press.
19. Thompson, C. J. and Coskuner-Balli, G. (2007). 'Countervailing Market Responses to Corporate Co-optation and the Ideological Recruitment of Consumption Communities'. *Journal of Consumer Research*, 34(2): 135–52; Loacker, B. (2013). 'Becoming "Culturpreneur": How the "Neoliberal Regime of Truth"

affects and Redefines Artistic Subject Positions'. *Culture and Organization*, 19(2): 124–45; Bain, A. and McLean, H. (2012). 'The Artistic Precariat'. *Cambridge Journal of Regions, Economy and Society*, 6(1): 93–111.

20. Mercier, A. (2010). 'Interview with Christian Arnsperger: Capitalism is Experiencing an Existential Crisis'. *Truthout*, 15 April.
21. Bloom, P. and Rhodes, C. (2018). *The CEO Society: The Corporate Takeover of Everyday Life*. Zed Books.
22. See Fries, C. J. (2008). 'Governing the Health of the Hybrid Self: Integrative Medicine, Neoliberalism, and the Shifting Biopolitics of Subjectivity'. *Health Sociology Review*, 17(4): 353–67; Young, S. L. and Reynolds Jr, D. A. J. (2017). '"You Can Be an Agent of Change": The Rhetoric of New Age Self-help in Enlightened'. *Western Journal of Communication*, 81(1): 1–20.
23. Doran, P. (2018). 'Mcmindfulness: Buddhism As Sold to You by Neoliberals'. *The Conversation*, 23 February.
24. Elliott, L. (2018). 'Robots Will Take Our Jobs: We'd Better Plan Now, Before It's Too Late'. *The Guardian*, 1 February.
25. See for instance Papadopoulos, D. and Tsianos, V. (2007). 'How to Do Sovereignty Without People? the Subjectless Condition of Postliberal Power'. *boundary 2*, 34(1): 135–72; Barnett, C., Clarke, N., Cloke, P. and Malpass, A. (2014). 'The Elusive Subjects of Neo-liberalism: Beyond the Analytics of Governmentality'. In Binkley, S. and Littler, J. (eds), *Cultural Studies and Anti-Consumerism*. Routledge.
26. Waihenya, W. (2018). 'Technology May Be Glorified But it Has Made Us Soulless Creatures'. *Daily Nation*, 15 June.
27. Frangos, J. M. (2014). 'Will Technology Rob Us of Our Humanity?'. *World Economic Forum*, 12 September.
28. Noto La Diega, G. (2018). 'Against the Dehumanisation of Decision-making: Algorithmic Decisions at the Crossroads of Intellectual Property, Data Protection, and Freedom of Information'. *Journal of Intellectual Property, Information Technology and Electronic Commerce Law*, 9(1).
29. O'Connor, S. (2016). 'When Your Boss is an Algorithm'. *Financial Times*, 8 September.
30. Harrison, M. (2015). 'Large Numbers Are Dehumanising, So Should Big Data Worry Us?'. *The Guardian*, 16 April.

31. Manovich, L. (2017). 'Cultural Analytics, Social Computing and Digital Humanities'. In Schaefer, M. and van Es, K. (eds), *The Datafied Society*: 55.
32. Cep. C. (2014). 'Big Data for the Soul'. *New Yorker*, 5 August.
33. See Rogers, R. (2017). 'Foundations of Digital Methods: Query Design'. In Schaefer, M. and van Es, K. (eds), *The Datafied Society*. Amsterdam University Press.
34. See Costas, J. and Fleming, P. (2009). 'Beyond Dis-identification: a Discursive Approach to Self-alienation in Contemporary Organizations'. *Human Relations*, 62(3): 353–78.
35. Abbott, A. (2000). 'Reflections on the Future of Sociology'. *Contemporary Sociology*, 29(2): 296–300.
36. Houtman, D. and Aupers, S. (2010). 'Religions of Modernity: Relocating the Sacred to the Self and the Digital'. In Aupers, S. and Houtman, D. (eds), *Religions of Modernity*. Brill: 6–7.
37. Grossman, C. (2014). 'New "Soulpulse" App Lets Users Monitor their Spirituality in Real Time'. *Washington Post*, 10 January.
38. Indick, W. (2015). *The Digital God: How Technology Will Reshape Spirituality*. McFarland.
39. Berry, D. M. (2012). 'Introduction: Understanding the Digital Humanities'. In Berry, D. M. (ed.), *Understanding Digital Humanities*. Palgrave Macmillan: 3.
40. Klauser, F. R. and Albrechtslund, A. (2014). 'From Self-tracking to Smart Urban Infrastructures: Towards an Interdisciplinary Research Agenda on Big Data'. Surveillance & Society, 12(2): 273.
41. Gray, K. (2017). 'Inside Silicon Valley's New Non-religion: Consciousness Hacking'. *Wired*, 1 November.
42. Dean, M. (2017). 'Political Acclamation, Social Media and the Public Mood'. *European Journal of Social Theory*, 20(3): 417–34.
43. Cobb, J. J. (1998). *Cybergrace: The Search for God in the Digital World*. Random House.
44. See for instance, Kanter, R. M. (1972). *Commitment and Community: Communes and Utopias in Sociological Perspective*, Volume 36). Harvard University Press.
45. See Luhmann, M. (2017). 'Using Big Data to Study Subjective Well-being'. *Current Opinion in Behavioral Sciences*, 18: 28–33.
46. Loewe, E. (2018). 'The Rise of Spiritual Tech'. *MBG Mindfulness*, 9 (February).

47. Krotoski, A. (2011). 'What Effect Has God Had on Religion'. *The Guardian*, 17 April.
48. Ibid.
49. Cederstrom, C. and Spicer, A. (2017). *Desperately Seeking Self-improvement: A Year Inside the Optimization Movement*. OR Books.
50. Marcengo, A. and Rapp, A. (2016). 'Visualization of Human Behavior Data: the Quantified Self'. In Information Resources Management Association, *Big Data: Concepts, Methodologies, Tools, and Applications*. IGI Global: 1582.
51. Hansen, M. (2000). *Embodying Technesis: Technology Beyond Writing*. University of Michigan Press: 17.
52. Stephens-Davidowitz, S. and Pinker, S. (2017). *Everybody Lies: Big Data, New Data, and What the Internet Can Tell Us About Who We Really Are*. HarperLuxe.
53. Wachter-Boettcher, S. (2017). *Technically Wrong: Sexist Apps, Biased Algorithms, and Other Threats of Toxic Tech*. W.W. Norton and Company.
54. Mathiesen, T. (1997). 'The Viewer Society: Michel Foucault's Panopticon Revisited'. *Theoretical Criminology*, 1(2): 215.
55. Davies, W. (2018). 'The Political Economy of Pulse: Techno-somatic Rhythm and Real-time Data'. *Ephemera*, 18(4).
56. James, G. 'How Steve Jobs Trained His Own Brain'. *Inc.com*.
57. See Widdicombe, L. (2015). 'The Higher Life'. *The New Yorker*, 6 July.
58. See McAbee, S. T., Landis, R. S. and Burke, M. I. (2017). 'Inductive Reasoning: The Promise of Big Data'. *Human Resource Management Review*, 27(2): 277–90.
59. Stephens-Davidowitz, S. (2017). 'Everybody Lies: How Google Search Reveals Our Darkest Secrets'. *The Guardian*, 9 July.
60. See Fox, J. and Tang, W. Y. (2014). 'Sexism in Online Video Games: The Role of Conformity to Masculine Norms and Social Dominance Orientation'. *Computers in Human Behavior*, 33, 314–20; Wolf, W., Levordashka, A., Ruff, J. R., Kraaijeveld, S., Lueckmann, J. M. and Williams, K. D. (2015). 'Ostracism Online: a Social Media Ostracism Paradigm'. *Behavior Research Methods*, 47(2): 361–73; Fuchs, C. (2015). *Culture and Economy in the Age of Social Media*. Routledge.
61. Pantzar, M. and Ruckenstein, M. (2015). 'The Heart of Everyday Analytics: Emotional, Material and Practical Extensions in Self-tracking Market'. *Consumption Markets & Culture*, 18(1): 92.

62. Ray, J. A. (2008). *Harmonic Wealth: The Secret of Attracting the Life You Want*. Hachette Books.
63. Foucault, M. (1982). 'The Subject and Power'. *Critical Inquiry*, 8(4): 783.
64. Schüll, N. D. (2016). 'Data for Life: Wearable Technology and the Design of Self-care'. *BioSocieties*, 11(3): 317.
65. Uricchio, W. (2017). 'Data, Culture and the Ambivalence of Algorithms'. In Schaefer, M. and van Es, K. (eds), *The Datafied Society*. Amsterdam University Press: 134.
66. Drucker, J. (2014). *Graphesis: Visual Forms of Knowledge Production*. Harvard University Press.
67. Stark, L. and Crawford, K. (2015). 'The Conservatism of Emoji: Work, Affect, and Communication'. *Social Media+ Society*, 1(2): 1–11.
68. Marwick, A. E. (2012). 'The Public Domain: Social Surveillance in Everyday Life'. Surveillance & Society, 9(4): 378.
69. Thatcher, J., O'Sullivan, D. and Mahmoudi, D. (2016). 'Data Colonialism Through Accumulation by Dispossession: New Metaphors for Daily Data'. *Environment and Planning D: Society and Space*, 34(6): 990.
70. See Bloom, P. (2016). 'Work As the Contemporary Limit of Life: Capitalism, the Death Drive, and the Lethal Fantasy of "Work–Life Balance"'. Organization, 23(4): 588–606.

Chapter 6

1. Kenna, R., MacCarron, M., and MacCarron, P. (eds) (2017). *Maths Meets Myths: Quantitative Approaches to Ancient Narratives*. Springer.
2. Gurrin, C., Smeaton, A. F. and Doherty, A. R. (2014). 'Lifelogging: Personal Big Data'. *Foundations and Trends in Information Retrieval*, 8(1): 2.
3. Mirowski, P. (2013). *Never Let a Serious Crisis Go to Waste: How Neoliberalism Survived the Financial Meltdown*. Verso.
4. De Cock, C., Vachhani, S. and Murray, J. (2013). 'Putting into Question the Imaginary of Recovery: A Dialectical Reading of the Global Financial Crisis and Its Aftermath'. *Culture and Organization*, 19(5): 396–412.

5. De Cock, C., Baker, M. and Volkmann, C. (2011). 'Financial Phantasmagoria: Corporate Image-work in Times of Crisis'. *Organization*, 18(2): 153–72.
6. Fukuyama, F. (2017). *The Great Disruption*. Profile Books.
7. Tegmark, M. (2017). *Life 3.0: Being Human in the Age of Artificial Intelligence*. Knopf.
8. Mokyr, J., Vickers, C. and Ziebarth, N. L. (2015). 'The History of Technological Anxiety and the Future of Economic Growth: Is This Time Different?'. *Journal of Economic Perspectives*, 29(3): 31–50.
9. Hirsch-Kreinsen, H., Weyer, J. and Wilkesmann, J. D. M. (2016). *'Industry 4.0' as Promising Technology: Emergence, Semantics and Ambivalent Character*. Universitätsbibliothek Dortmund.
10. Ayres, E. (2018). Defying Dystopia: Going on with the Human Journey After Technology Fails Us. Taylor and Francis.
11. Shackle, G. L. S. (1961). *Decision, Order, and Time in Human Affairs*, Volume 2. Cambridge University Press: 10.
12. Bloom, P. (2018) *The Bad Faith in the Free Market: The Radical Promise of Existential Freedom*. Palgrave Macmillan.
13. Rovelli, C. (2018). 'The Order of Time'. *C-span*.
14. Giroux, H. A. (2011). *Zombie Politics and Culture in the Age of Casino Capitalism*. Peter Lang.
15. Lauro, S. J. and Embry, K. (2008). 'A Zombie Manifesto: The Nonhuman Condition in the Era of Advanced Capitalism'. *boundary 2*, 35(1): 85–108; Lanci, Y. (2014). 'Zombie 2.0: Subjectivation in Times of Apocalypse'. *Journal for Cultural and Religious Theory*, 13(2): 25–37. For the more radical possibilities of this 'zombie culture' see Schneider, R. (2012). 'It Seems As If … I Am Dead: Zombie Capitalism and Theatrical Labor'. *TDR/The Drama Review*, 56(4): 150–62.
16. See Noys, B. (2014). *Malign Velocities: Accelerationism and Capitalism*. John Hunt Publishing.
17. Williams, A. and Srnicek, N. (2013). '#Accelerate: Manifesto for an Accelerationist Politics'. *Critical Legal Thinking*, 14 May.
18. See Noys, B. (2013). 'Days of Phuture Past: Accelerationism in the Present Moment'. Unpublished paper; O'Sullivan, S. (2015). *Accelerationism, Hyperstition and Myth-science: Accelerationism and the Occult*. Punctum Books.
19. Kunkel, B. (2008). 'Dystopia and the End of Politics'. *Dissent*, 55(4): 89–98.

20. Comaroff, J. and Comaroff, J. L. (2001). 'Naturing the Nation: Aliens, Apocalypse, and the Postcolonial State'. *Social Identities*, 7(2): 233–65.
21. See Swyngedouw, E. (2010). 'Apocalypse Forever?'. *Theory, Culture and Society*, 27(2–3): 213–32.
22. Leyshon, A. and Thrift, N. (2007). 'The Capitalization of Almost Everything: The Future of Finance and Capitalism'. *Theory, Culture & Society*, 24(7–8): 97–115.
23. Crary, J. (2013). *24/7: Late Capitalism and the Ends of Sleep*. Verso. Also see Hassan, R. and Purser, R. E. (2007). *24/7: Time and Temporality in the Network Society*. Stanford University Press.
24. Graham, C. (2012). 'The Subject of Retirement'. *Foucault Studies*, 13: 25–39.
25. Neilson, B. and Rossiter, N. (2005). 'From Precarity to Precariousness and Back Again: Labour, Life and Unstable Networks'. *Fibreculture*, 5.
26. See Rudman, D. L. (2005). 'Understanding Political Influences on Occupational Possibilities: An Analysis of Newspaper Constructions of Retirement'. *Journal of Occupational Science*, 12(3): 149–60.
27. Polivka, L. (2011). 'Neoliberalism and Postmodern Cultures of Aging'. *Journal of Applied Gerontology*, 30(2): 173–84.
28. Tietz, T., Pichierri, F., Koutraki, M., Hallinan, D., Boehm, F. and Sack, H. (2018). 'Digital Zombies: The Reanimation of our Digital Selves'. In International World Wide Web Conferences Steering Committee, *Companion Proceedings of the Web Conference 2018*.
29. Ibid. See for example the episode where this is done with quite devastating results in an episode of the Netflix show *Black Mirror*.
30. Gould, P. (1981). 'Letting the Data Speak for Themselves'. *Annals of the Association of American Geographers*, 71(2): 166–76.
31. Parry, O. and Mauthner, N. S. (2004). 'Whose Data Are they Anyway? Practical, Legal and Ethical Issues in Archiving Qualitative Research Data'. *Sociology*, 38(1): 139–52.
32. Pötzsch, H. (2017). 'Archives and Identity in the Context of Social Media and Algorithmic Analytics: Towards an Understanding of iArchive and Predictive Retention'. *New Media & Society*, online first.

33. Burkell, J. A. (2016). 'Remembering Me: Big Data, Individual Identity, and the Psychological Necessity of Forgetting'. *Ethics and Information Technology*, 18(1): 17–23.
34. See Przybylski, A. K., Murayama, K., DeHaan, C. R. and Gladwell, V. (2013). 'Motivational, Emotional, and Behavioral Correlates of Fear of Missing Out'. *Computers in Human Behavior*, 29(4): 1841–8.
35. Latour, B. (1987). *Science in Action: How to Follow Scientists and Engineers Through Society*. Harvard University Press: 224.
36. Robson, K. (1992). 'Accounting Numbers as "Inscription": Action at a Distance and the Development of Accounting'. *Accounting, Organizations and Society*, 17(7): 685–708.
37. Lemov, R. (2015). *Database of Dreams: The Lost Quest to Catalog Humanity*. Yale University Press.
38. Lemov, R. (2016) 'Big Data is People'. *Aeon*, 16 June.
39. Callon, M. (2007). 'Some Elements of a Sociology of Translation'. In Asdal, K., Brenna, B. and Moser I. (eds), *Technoscience: The Politics of Interventions*. Unipub: 65.
40. See Dambrin, C. and Robson, K. (2010). *Multiple Measures, Inscription Instability and Action At a Distance: Performance Measurement Practices in the Pharmaceutical Industry*. Groupe HEC.
41. Miller, P. and Rose, N. (2008). Governing the Present: Administering Economic, Social and Personal Life. Polity: 18.
42. See for instance Makovicky, N. (2016). Neoliberalism, Personhood, and Postsocialism: Enterprising Selves in Changing Economies. Routledge.
43. Lupton, D. (2016). 'You Are Your Data: Self-tracking Practices and Concepts of Data'. In Selke , S. (ed.), *Lifelogging*. Springer: 61.
44. Fayyad, U., Piatetsky-Shapiro, G. and Smyth, P. (1996). 'From Data Mining to Knowledge Discovery in Databases'. *AI Magazine*, 17(3). Also see Han, J., Pei, J. and Kamber, M. (2011). *Data Mining: Concepts and Techniques*. Elsevier. Larose, D. T. and Larose, C. D. (2014). Discovering Knowledge in Data: An Introduction to Data Mining. John Wiley and Sons.
45. Tirunillai, S. and Tellis, G. J. (2014). 'Mining Marketing Meaning from Online Chatter: Strategic Brand Analysis of Big Data Using Latent Dirichlet Allocation'. *Journal of Marketing Research*, 51(4): 463–79.

46. Caruso, C., Dimitri, A. and Mecella, M. (2016). 'Identity Mining vs Identity Discovering: A New Approach'. Unpublished manuscript.
47. Tsikerdekis, M. (2017). 'Real-time Identity Deception Detection Techniques for Social Media: Optimizations and Challenges'. *IEEE Internet Computing*.
48. Velandia, D. M. S., Kaur, N., Whittow, W. G., Conway, P. P. and West, A. A. (2016). 'Towards Industrial Internet of Things: Crankshaft Monitoring, Traceability and Tracking Using RFID'. *Robotics and Computer-integrated Manufacturing*, 41, 66–77.
49. See Patel, J. (2016). 'Real Time Big Data Mining'. Doctoral dissertation, Rutgers University-Camden Graduate School; Zheng, Z., Wang, P., Liu, J. and Sun, S. (2015). 'Real-time Big Data Processing Framework: Challenges and Solutions'. *Applied Mathematics & Information Sciences*, 9(6): 3169.
50. Beer, D. (2017). 'The Data Analytics Industry and the Promises of Real-time Knowing: Perpetuating and Deploying a Rationality of Speed'. *Journal of Cultural Economy*, 10(1): 21.
51. Baruh, L. and Popescu, M. (2017). 'Big Data Analytics and the Limits of Privacy Self-management'. *New Media & Society*, 19(4): 597.
52. Jacobs, A. (2009). 'The Pathologies of Big Data'. *Communications of the ACM*, 52(8): 36–44.
53. Hutchins, B. (2016). 'Tales of the Digital Sublime: Tracing the Relationship Between Big Data and Professional Sport'. *Convergence*, 22(5): 494–509.
54. Kwan, M. P. (2016). 'Algorithmic Geographies: Big Data, Algorithmic Uncertainty, and the Production of Geographic Knowledge'. *Annals of the American Association of Geographers*, 106(2): 274–82.
55. Dalton, C. M., Taylor, L. and Thatcher, J. (2016). 'Critical Data Studies: A Dialog on Data and Space'. *Big Data & Society*, 3(1): 6.
56. Haggerty, K. D. and Ericson, R. V. (2000). 'The Surveillant Assemblage'. *The British Journal of Sociology*, 51(4): 605–22.
57. Obermeyer, Z. and Emanuel, E. J. (2016). 'Predicting the Future: Big Data, Machine Learning, and Clinical Medicine'. *The New England Journal of Medicine*, 375(13): 1216.
58. See Asur, S. and Huberman, B. A. (2010). 'Predicting the Future with Social Media'. In IEEE Computer Society, *Proceedings of the 2010 IEEE/WIC/ACM International Conference on Web Intelligence*

and Intelligent Agent Technology, Volume 1; Choi, H. and Varian, H. (2012). 'Predicting the Present with Google Trends'. *Economic Record*, 88: 2–9.

59. Gilbert, D. T., Gill, M. J. and Wilson, T. D. (2002). 'The Future is Now: Temporal Correction in Affective Forecasting'. *Organizational Behavior and Human Decision Processes*, 88(1): 430–44.
60. Martin, J. (2012). 'Second Life Surveillance: Power to the People or Virtual Surveillance Society?'. *Surveillance & Society*, 9(4): 408.
61. Verma, N. and Voida, A. (2016). 'On Being Actionable: Mythologies of Business Intelligence and Disconnects in Drill Downs'. In *Proceedings of the 19th International Conference on Supporting Group Work*: 35.
62. Howell, N., Devendorf, L., Vega Gálvez, T. A., Tian, R. and Ryokai, K. (2018). 'Tensions of Data-Driven Reflection: A Case Study of Real-time Emotional Biosensing'. In *Proceedings of the 2018 CHI Conference on Human Factors in Computing Systems*. ACM.
63. Bury, J. B. (1987). *The Idea of Progress: An Inquiry into Its Origin and Growth*. Courier Corporation.
64. Nisbet, R. (2017). *History of the Idea of Progress*. Routledge: chapter 9 abstract.
65. Allen, A. (2016). *The End of Progress: Decolonizing the Normative Foundations of Critical Theory*. Columbia University Press.
66. McMahon, J. (2015). 'Behavioral Economics As Neoliberalism: Producing and Governing Homo Economicus'. *Contemporary Political Theory*, 14(2): 137–58.
67. Brynjolfsson, E. (2016) 'The Rise of Data Capital'. MIT and Oracle Report.
68. Kitchin, R. (2014). *The Data Revolution: Big Data, Open Data, Data Infrastructures and their Consequences*. Sage.
69. Mai, J. E. (2016). 'Big Data Privacy: The Datafication of Personal Information'. *The Information Society*, 32(3): 192.
70. Safier, S. (2000). 'Between Big Brother and the Bottom Line: Privacy in Cyberspace'. *Virginia Journal of Law & Technology*, 5: 1.
71. Pentland, A. (2009). 'Reality Mining of Mobile Communications: Toward a New Deal on Data'. In Dutta, S. and Mia, I. (eds), *The Global Information Technology Report 2008–2009*. World Economic Forum.

72. See Eagle, N. and Pentland, A. S. (2006). 'Reality Mining: Sensing Complex Social Systems'. *Personal and Ubiquitous Computing*, 10(4): 255–68.
73. Pentland, A. (2009). 'Reality Mining of Mobile Communications: Toward a New Deal on Data'. In Dutta, S. and Mia, I. (eds), *The Global Information Technology Report 2008–2009*. World Economic Forum: 80.
74. Reigeluth, T. (2014). 'Why Data is Not Enough: Digital Traces As Control of Self and Self-Control'. *Surveillance & Society*, 12(2): 251.
75. Taylor, L. and Schroeder, R. (2015). 'Is Bigger Better? the Emergence of Big Data As a Tool for International Development Policy'. *GeoJournal*, 80(4): 503–18.
76. Thatcher, J. (2014). 'Big Data, Big Questions / Living on Fumes: Digital Footprints, Data Fumes, and the Limitations of Spatial Big Data'. *International Journal of Communication*, 8: 19.
77. Coté, M. (2014). 'Data Motility: The Materiality of Big Social Data'. *Cultural Studies Review*, 20(1): 123.
78. Amoore, L. and Piotukh, V. (2015). 'Life Beyond Big Data: Governing with Little Analytics'. *Economy and Society*, 44(3): 341–66.
79. Selwyn, N. (2015). 'Data Entry: Towards the Critical Study of Digital Data and Education'. *Learning, Media and Technology*, 40(1): 64–82.
80. Cukier, K. and Mayer-Schoenberger, V. (2013). 'The Rise of Big Data: How It's Changing the Way we Think About the World'. *Foreign Affairs*, 92: 28–40.
81. Beer, D. (2017). 'The Social Power of Algorithms'. *Information, Communication and Society*, 20(1): 1.
82. Crawford, K. (2016). 'The Anxieties of Big Data'. *New Inquiry*, 30 May.
83. Ibid.
84. Leszczynski, A. (2015). 'Spatial Big Data and Anxieties of Control'. *Environment and Planning D: Society and Space*, 33(6): 965–84.
85. Hilbert, M. (2013). 'Big Data for Development: From Information-to Knowledge Societies'. Unpublished paper.
86. Kaplan, M. (2018). '"Spying for the People": Surveillance, Democracy and the Impasse of Cynical Reason'. *JOMEC Journal*, 12: 166–90.

Chapter 7

1. Botsman, R. (2017) 'Big Data Meets Big Brother As China Moves to Rate Its Citizens'. *Wired*, 21 October.
2. CCP (2015) 'Planning Outline for the Construction of a Social Credit System (2014–2020)'. Updated 25 April.
3. See Bad Social Credit score, Fullerton, J. (2018). 'China's "Social Credit" System Bans Millions from Travelling'. *The Telegraph*, 24 March; Mistreanu, S. (2018). 'Life Inside China's Social Credit Laboratory'. *Foreign Policy*, 3; Brown, J. (2018). 'Would You Choose a Partner Based on their "Citizen Score"?'. BBC, 13 March.
4. Ma, A. (2018). 'China Ranks Citizens with a Social Credit System: Here's What You Can Do Wrong and How You Can Be Punished'. *The Independent*, 10 April.
5. Zeng, M. (2018). 'China's Social Credit System Puts Its People Under Pressure to Be Model Citizens'. *The Conversation*, 23 January.
6. Greenfield, A. (2018). 'China's Dystopian Tech Could be Contagious'. *The Atlantic*, 14 February.
7. Zeng, M. (2018). 'China's Social Credit System Puts Its People Under Pressure to Be Model Citizens'. *The Conversation*, 23 January.
8. Mayer-Schönberger, V. and Cukier, K. (2013). *Big Data: A Revolution that Will Transform How We Live, Work, and Think*. John Murray.
9. Corera, G. (2015). 'Will Big Data Lead to Big Brother?'. BBC, 17 November.
10. Lesk, M. (2013). 'Big Data, Big Brother, Big Money'. *IEEE Security & Privacy*, 11(4): 85–9.
11. See Ziewitz, M. (2016). 'Governing Algorithms: Myth, Mess, and Methods'. *Science, Technology, & Human Values*, 41(1): 3–16.
12. See Musiani, F. (2013). 'La gouvernance des algorithmes'. Hypotheses.org.
13. Kitchin, R. (2013). 'Big Data and Human Geography: Opportunities, Challenges and Risks'. *Dialogues in Human Geography*, 3(3): 262–7.
14. Estrin, D. (2014) 'Small Data, Where N = Me'. *Communications of the ACM*, April: 32.

15. See Girardin, F., Calabrese, F., Dal Fiore, F., Ratti, C. and Blat, J. (2008). 'Digital Footprinting: Uncovering Tourists with User-generated Content'. *IEEE Pervasive Computing*, 7(4).
16. Dalton, C. M. and Thatcher, J. (2015). 'Inflated Granularity: Spatial "Big Data" and Geodemographics'. *Big Data & Society*, 2(2).
17. See Mann, S., Nolan, J. and Wellman, B. (2003). 'Sousveillance: Inventing and Using Wearable Computing Devices for Data Collection in Surveillance Environments'. *Surveillance & Society*, 1(3): 331–55.
18. Mann, S., Ali, M. A., Lo, R. and Wu, H. (2013). 'Freeglass for Developers' Haccessibility and Digital Eye Glass+ Lifeglogging Research in a (Sur/Sous) Veillance society'. In IEEE, *Information Society (i-Society): 2013 International Conference.*
19. Perez, C. (2017). 'How Algorithms and Authoritarianism Created a Corporate Nightmare at United'. *Medium*, 14 April.
20. Foster, P. (2018) 'Why Big Data is Killing Western Democracy: And Giving Authoritarian States a New Lease of Life'. *The Telegraph*, 24 April.
21. 'The Big Data Revolution Can Revive the Planned Economy'. *Financial Times*, 17 September.
22. See Bloom, P. (2016). *Authoritarian Capitalism in the Age of Globalization*. Edward Elgar.
23. See Williamson, B. (2016). 'Digital Education Governance: Data Visualization, Predictive Analytics, and "Real-time" Policy Instruments'. *Journal of Education Policy*, 31(2): 123–41.
24. Hansen, H. K. and Flyverbom, M. (2015). 'The Politics of Transparency and the Calibration of Knowledge in the Digital Age'. *Organization*, 22(6): 872.
25. See Schrock, A. and Shaffer, G. (2017). 'Data Ideologies of an Interested Public: A Study of Grassroots Open Government Data Intermediaries'. *Big Data & Society*, 4(1).
26. See Ashley, M. (2009). 'Deep Thinking in Shallow Time: Sharing Humanity's History in the Petabyte Age'. In Tsipopoulou, M. (ed.), *Proceedings for Digital Heritage in the New Knowledge Environment: Shared Spaces and Open Paths to Cultural Content*; Brand, S. (1999). 'Escaping the Digital Dark Age'. *Library Journal*, 124(2): 46–8.

27. Hargittai, E. (2015). 'Is Bigger Always Better? Potential Biases of Big Data Derived from Social Network Sites'. *The ANNALS of the American Academy of Political and Social Science*, 659(1): 63–76.
28. See especially Noble, S. U. (2018). *Algorithms of Oppression: How Search Engines Reinforce Racism*. NYU Press.
29. Madden, M., Gilman, M., Levy, K., and Marwick, A. (2017). 'Privacy, Poverty, and Big Data: A Matrix of Vulnerabilities for Poor Americans'. *Washington University Law Review*, 95(2): 53.
30. Cyril, M. (2017). 'The Antidote to Technology'. *The Atlantic*, 8 May.
31. Andrejevic, M. and Gates, K. (2014). 'Big Data Surveillance: Introduction'. *Surveillance & Society*, 12(2): 192.
32. O'Neil, C. (2016). 'Weapons of Math Destruction: How Big Data Increases Inequality and Threatens Democracy'. Broadway Books.
33. Osborne, P. (1995). *The Politics of Time*. Verso: 196.
34. Cohen, E. F. (2018). *The Political Value of Time: Citizenship, Duration, and Democratic Justice*. Cambridge University Press.
35. Oliver, N. (2018). 'The Tyranny of Data? The Bright and Dark Sides of Algorithmic Decision Making for Public Policy Making'. In Gómez, E. (ed.), *Assessing the Impact of Machine Intelligence on Human Behaviour: An Interdisciplinary Endeavour*. European Commission: 58.
36. Chandler, D. (2015). 'A World Without Causation: Big Data and the Coming of Age of Posthumanism'. *Millennium*, 43(3): 833–51.
37. Kitchin, R. (2014). 'The Real-time City? Big Data and Smart Urbanism'. *GeoJournal*, 79(1): 1.
38. Shelton, T. and Clark, J. (2016). 'Technocratic Values and Uneven Development in the "Smart City"'. *Metropolitics/Metropolitiques*.
39. See Reich, R. (1991). *The Wealth of Nations*. Alfred A. Knopf.
40. Van Otterlo, M. (2014). 'Automated Experimentation in Walden 3.0: The Next Step in Profiling, Predicting, Control and Surveillance'. *Surveillance & Society*, 12(2): 255.
41. See for instance Grosser, B. (2017). 'Tracing You: How Transparent Surveillance Reveals a Desire for Visibility'. *Big Data & Society*, 4(1).
42. Wajcman, J. (2015). *Pressed for Time: The Acceleration of Life in Digital Capitalism*. University of Chicago Press.
43. Fuchs, C. (2014). 'Digital Prosumption Labour on Social Media in the Context of the Capitalist Regime of Time'. *Time & Society*, 23(1): 97–123.

44. Fleming, P. and Spicer, A. (2004). '"You Can Checkout Anytime, But You Can Never Leave": Spatial Boundaries In A High Commitment Organization'. *Human Relations*, 57(1): 75–94.
45. Whitaker, R. (2010). *The End of Privacy: How Total Surveillance Is Becoming a Reality*. ReadHowYouWant.Com.
46. Leistert, O. (2012). 'Resistance Against Cyber-surveillance Within Social Movements and How Surveillance Adapts'. *Surveillance & Society*, 9(4): 441.
47. Rouvroy, A. (2013). 'The End(s) of Critique: Data Behaviourism Versus Due Process'. In Hildebrandt, M. and de Vries, K. (eds), *Privacy, Due Process and the Computational Turn*. Routledge: 147.
48. Williamson, B. (2018). 'Silicon Startup Schools: Technocracy, Algorithmic Imaginaries and Venture Philanthropy in Corporate Education Reform'. *Critical Studies in Education*, 59(2): 218.
49. Christin, A. (2016). 'From Daguerreotypes to Algorithms: Machines, Expertise, and Three Forms of Objectivity'. *ACM SIGCAS Computers and Society*, 46(1): 27–32.
50. See Kosinski, M., Stillwell, D. and Graepel, T. (2013). 'Private Traits and Attributes Are Predictable from Digital Records of Human Behavior'. *Proceedings of the National Academy of Sciences*, 110(15): 5802–5.
51. See Bloom, P. (2017). *The Ethics of Neoliberalism: The Business of Making Capitalism Moral*. Routledge.
52. Moore, P. and Robinson, A. (2016). 'The Quantified Self: What Counts in the Neoliberal Workplace'. *New Media & Society*, 18(11): 2774–92.
53. Wilson, K. (2013). 'Agency as "Smart Economics": Neoliberalism, Gender and Development'. In Madhok, S., Phillips, A., Wilson, K. and Hemmings, C. (eds), *Gender, Agency, and Coercion*. Palgrave Macmillan.
54. Sanders, R. (2017). 'Self-tracking in the Digital Era: Biopower, Patriarchy, and the New Biometric Body Projects'. *Body & Society*, 23(1): 36–63.
55. See Gehl, R. W. and Bakardjieva, M. (eds) (2016). *Socialbots and their Friends: Digital Media and the Automation of Sociality*. Taylor and Francis.
56. Munger, K. (2016) 'This Researcher Programmed Bots to Fight Racism on Twitter: It Worked'. *Washington Post*, 17 November.
57. See Jee, C. (2018). 'The Best Ways Businesses Are Using Chatbots'. *Techworld*, 10 April.

58. See Chan, J. and Bennett Moses, L. (2017). 'Making Sense of Big Data for Security'. *The British Journal of Criminology*, 57(2): 299–319; Casciani, D. (2015). 'Three Ways Secret Data Collection Fights Crime'. BBC, 11 June.
59. Bain, P. and Taylor, P. (2000). 'Entrapped by the "Electronic Panopticon"? Worker Resistance in the Call Centre'. *New Technology, Work and Employment*, 15(1): 2–18.
60. Jackson, P., Gharavi, H. and Klobas, J. (2006). 'Technologies of the Self: Virtual Work and the Inner Panopticon'. *Information Technology & People* 19(3): 219–43.
61. See Rosenfeld, P., Booth-Kewley, S., Edwards, J. E. and Thomas, M. D. (1996). 'Responses on Computer Surveys: Impression Management, Social Desirability, and the Big Brother Syndrome'. *Computers in Human Behavior*, 12(2): 263–74.
62. Attewell, P. (1987). 'Big Brother and the Sweatshop: Computer Surveillance in the Automated Office'. *Sociological Theory*, 87–100.
63. DeTienne, K. B. (1993). 'Big Brother or Friendly Coach? Computer Monitoring in the 21st Century'. *The Futurist*, 27(5): 33.
64. See Trottier, D. (2016). *Social Media As Surveillance: Rethinking Visibility in a Converging World*. Routledge; Norris, C. and Moran, J. (2016). *Surveillance, Closed Circuit Television and Social Control*. Routledge.
65. Anon. 'DHS Science And Technology Directorate: Future Attribute Screening Technology'. Homeland Security. Electronic resource: www.dhs.gov/sites/default/files/publications/Future%20Attribute%20Screening%20Technology-FAST.pdf.
66. Harper, T. (2017). 'The Big Data Public and Its Problems: Big Data and the Structural Transformation of the Public Sphere'. *New Media & Society*, 19(9): 1424.
67. Yoo, Y., Boland Jr, R. J., Lyytinen, K. and Majchrzak, A. (2012). 'Organizing for Innovation in the Digitized World'. *Organization Science*, 23(5): 1398.
68. Just, N. and Latzer, M. (2017). 'Governance by Algorithms: Reality Construction by Algorithmic Selection on the Internet'. *Media, Culture & Society*, 39(2): 238.
69. Gilbert, J. (2013). 'What Kind of Thing is "Neoliberalism"?'. *New Formations: A Journal of Culture/Theory/Politics*, 80(80): 7.

70. Kitchin, R. and Lauriault, T. P. (2014). 'Towards Critical Data Studies: Charting and Unpacking Data Assemblages and their Work'. The Programmable City Working Paper, 2.
71. Anderson, C. (2008). 'The End of Theory: the Data Deluge Makes the Scientific Method Obsolete'. *Wired*, 16 July.
72. Guston, D. H. (2008). 'Preface'. In Fisher, E., Selin, C. and Wetmore, J. M. (eds), *The Yearbook of Nanotechnology in Society: Presenting Futures*, Vol. 1. Springer: v–viii.
73. See Milakovich, M. (2012). 'Anticipatory Government: Integrating Big Data for Smaller Government'. Paper for conference Internet, Politics, Policy 2012: Big Data, Big Challenges; Evans, K. G. (1997). 'Imagining Anticipatory Government: A Speculative Essay on Quantum Theory and Visualization'. *Administrative Theory & Praxis*, 355–67.
74. Guston, D. H. (2014). 'Understanding "Anticipatory Governance"'. *Social Studies of Science*, 44(2): 234.
75. Jones, C. and Spicer, A. (2005). 'The Sublime Object of Entrepreneurship'. *Organization*, 12(2): 237.
76. Fuller, S. (2010). 'The New Behemoth (Review of "Acting in an Uncertain World")'. *Contemporary Sociology*, 39(5): 533–6.
77. Bloom, P. and Rhodes, C. (2018). *The CEO Society*. University of Chicago Press.

Chapter 8

1. See Lyon, D. (2014). 'Surveillance, Snowden, and Big Data: Capacities, Consequences, Critique'. *Big Data & Society*, 1(2); Lyon, D. (2016). 'Big Data Surveillance: Snowden, Everyday Practices and Digital Futures'. In Basaran, T., Bigo, D., Guittet, E.-P. and Walker, R. B. J. (eds), *International Political Sociology*. Routledge.
2. Mann, S. and Ferenbok, J. (2013). 'New Media and the Power Politics of Sousveillance in a Surveillance-dominated World'. *Surveillance & Society*, 11(1/2): 18–34.
3. Kennedy, H. and Moss, G. (2015). 'Known or Knowing Publics? Social Media Data Mining and the Question of Public Agency'. *Big Data & Society*, 2(2): 1.
4. Papacharissi, Z. (2015). 'The Unbearable Lightness of Information and the Impossible Gravitas of Knowledge: Big Data and the

Makings of a Digital Orality'. *Media, Culture & Society*, 37(7): 1095–100.

5. Mansell, R. (2017). 'Imaginaries, Values, and Trajectories: A Critical Reflection on the Internet'. In Goggin, G. and McLelland, M. (eds), *The Routledge Companion to Global Internet Histories*. Routledge.
6. See Boyd, E., Nykvist, B., Borgström, S. and Stacewicz, I. A. (2015). 'Anticipatory Governance for Social-ecological Resilience'. *Ambio*, 44(1): 149–61.
7. Vlachokyriakos, V., Crivellaro, C., Le Dantec, C. A., Gordon, E., Wright, P. and Olivier, P. (2016). 'Digital Civics: Citizen Empowerment With and Through Technology'. In *Proceedings of the 2016 CHI Conference Extended Abstracts on Human Factors in Computing Systems*: 1096.
8. Tapscott, D. and Tapscott, A. (2016). *Blockchain Revolution: How the Technology Behind Bitcoin is Changing Money, Business, and the World*. Penguin.
9. Heikka, T. (2015). 'The Rise of the Mediating Citizen: Time, Space, and Citizenship in the Crowdsourcing of Finnish Legislation'. *Policy & Internet*, 7(3): 268–91.
10. See DeLyser, D. and Sui, D. (2014). 'Crossing the Qualitative–Quantitative Chasm III: Enduring Methods, Open Geography, Participatory Research, and the Fourth Paradigm'. *Progress in Human Geography*, 38(2): 294–307.
11. Larner, J. and Mullagh, L. (2017). 'Critical Reflection on Data Publics: a Multi-methodology Perspective'. *Data Publics*.
12. Baack, S. (2015). 'Datafication and Empowerment: How the Open Data Movement Re-articulates Notions of Democracy, Participation, and Journalism'. *Big Data & Society*, 2(2): 1.
13. Mehta, P. (2015) 'Big Data's Radical Potential'. *Jacobin*, 12 March.
14. Howell, N. and Niemeyer, G. (2018). 'Reconfiguring Desires and Data'. Report from University California, Berkeley, March.
15. Ibid.
16. Foucault, M. (1982). 'The Subject and Power'. *Critical Inquiry*, 8(4): 788.
17. Foucault, M. (2013). *Archaeology of Knowledge*. Routledge: 40.
18. See Yang, S. H. (2009). 'Using Blogs to Enhance Critical Reflection and Community of Practice'. *Journal of Educational Technology & Society*, 12(2).

19. Barros, M. (2018). 'Digitally Crafting a Resistant Professional Identity: the Case of Brazilian "Dirty" Bloggers'. *Organization*, https://doi.org/10.1177/1350508418759185.
20. Campbell, P. (2015). *Digital Selves: Iraqi Women's Warblogs and the Limits of Freedom*. Common Ground: xxv.
21. Couldry, N. and Powell, A. (2014). 'Big Data from the Bottom Up'. *Big Data & Society*, 1(2): 1–5.
22. See Radsch, C. (2012). 'Unveiling the Revolutionaries: Cyberactivism and the Role of Women in the Arab Uprisings'. James A. Baker III Institute for Public Policy, Rice University.
23. See Callahan, M. (2015). Review of Todd Wolfson, *Digital Rebellion: The Birth of the Cyber Left*. The Research Group on Socialism and Democracy.
24. Gamie, S. (2013). 'The Cyber-propelled Egyptian Revolution and the De/Construction of Ethos'. In Folk, M. and Apostel, S. (eds), *Online Credibility and Digital Ethos: Evaluating Computer-mediated Communication*. IGI Global.
25. Brown, A. M. and Imarisha, W. (eds) (2015). *Octavia's Brood: Science Fiction Stories from Social Justice Movements*. AK Press: 3.
26. See Heide, M. (2015). 'Social Intranets and Internal Communication: Dreaming of Democracy in Organisations'. In Coombs, W. T. Falkheimer, J., Heide, M. and Young, P. (eds), *Strategic Communication, Social Media and Democracy*. Routledge.
27. See Jiménez, L. F. (2015). 'The Dictatorship Game: Simulating a Transition to Democracy'. *PS: Political Science & Politics*, 48(2): 353–7.
28. Selke, S. (ed.) (2016). *Lifelogging: Digital Self-tracking and Lifelogging: Between Disruptive Technology and Cultural Transformation*. Springer.
29. Ganguly, A., Nilchiani, R. and Farr, J. V. (2017). 'Technology Assessment: Managing Risks for Disruptive Technologies'. In Daim T. U. (ed.), *Managing Technological Innovation: Tools and Methods*. World Scientific.
30. Kroker, A. and Weinstein, M. A. (2015). *The Political Economy of Virtual Reality: Pan-capitalism*. Ctheory.
31. Romele, A. and Severo, M. (2016). 'The Economy of the Digital Gift: From Socialism to Sociality Online'. *Theory, Culture & Society*, 33(5): 43–63.
32. Davies, W. (ed.) (2018). *Economic Science Fictions*. MIT Press: xii.

33. Hughes, J. (2004). *Citizen Cyborg: Why Democratic Societies Must Respond to the Redesigned Human of the Future.* Basic Books.
34. Gray, C. H. (2000). *Cyborg Citizen: Politics in the Posthuman Age.* Routledge.
35. Koch, A. (2005). 'Cyber Citizen or Cyborg Citizen: Baudrillard, Political Agency, and the Commons in Virtual Politics'. *Journal of Mass Media Ethics*, 20(2–3): 159–75.
36. Gray, C. H. and Gordo, Á. J. (2014). 'Social Media in Conflict: Comparing Military and Social-movement Technocultures'. *Cultural Politics*, 10(3): 251.
37. Ibid.
38. Townsend, A. M. (2013). *Smart Cities: Big Data, Civic Hackers, and the Quest for a New Utopia.* W.W. Norton and Company.
39. See König, T. (2014). 'Revolutionaries' Tech Support: Hacktivism and Anonymous in the Egyptian Uprising'. In Hamed, A. (ed.), *Revolution as a Process: The Case of the Egyptian Uprising.* Wiener Verlag für Sozialforschung; Also Soldatov, A. and Borogan, I. (2015). *The Red Web: the Struggle Between Russia's Digital Dictators and the New Online Revolutionaries.* Hachette.
40. Neff, G. (2013). 'Why Big Data Won't Cure Us'. *Big Data*, 1(3): 117–23.
41. Halpern, O. (2015). 'The Trauma Machine: Demos, Immersive Technologies and the Politics of Simulation'. In Pasquinelli, M. (ed.), *Alleys of Your Mind: Augmented Intellligence and Its Traumas.* Meson Press.
42. Lupton, D. (2015). 'Quantified Sex: A Critical Analysis of Sexual and Reproductive Self-tracking Using Apps'. *Culture, Health & Sexuality*, 17(4): 440–53.
43. Stavrakakis, Y. (2007). *Lacanian Left.* Edinburgh University Press. Also see Glynos, J. (2000). 'Thinking the Ethics of the Political in the Context of a Postfoundational World: From an Ethics of Desire to an Ethics of the Drive'. *Theory & Event*, 4(4).

Index

Accelerationism 141
Actor Network Theory (ANT) 92
Algorithm vii, 9, 20, 26, 35, 39, 42, 50, 59, 80, 120, 123, 127, 128, 130, 132, 138, 140, 144, 145, 148–71, 175–81, 192–4
Amazon 27,48, 52, 183
Archiving ix, 56, 64, 68, 80, 140, 161
Assange, Julian 187
Auditing ix, 76
Authoritarianism 164–6, 169, 175, 183
Avatars 12, 52

Berry, David 122
Bauman, Zygmunt 54, 83, 110
  Liquid modernity 54–5, 83
Big Brother vii, 4–5, 156, 162, 164, 176, 178, 179, 182, 185, 187, 202
Blogs 52, 126, 193
  Iraqi Women 'war blogs' 193
Boltanski, Luc 105, 116
  New Spirit of Capitalism 105, 116

Cambridge Analytica 1–3, 163
Capitalism
  24/7 101, 142, 170
  Data 33–5, 197, 200
  Disaggregated 45, 183
  New Age 129
Castells, Manuel 57
Cederstrom, Carl 86, 112, 126
  Optimisation 126
  The Wellness Syndrome 112
Chiapello, Eve 105, 116
  New Spirit of Capitalism 105, 116
China 35, 162–3
The Circle vii, x, 14
The Cold War 138
Colonialism 17, 88–9, 102–7, 111, 132
Crawford, Kate 159

Data
  Addiction 41
  Arms Race 148, 169
  Assemblages 41, 47, 182
  Big Data vi–xiv, 2–9, 14, 20, 24–7, 30–47, 56, 67, 70, 77–8, 83, 87–9, 91, 94, 120–3, 126–7, 131–2, 138–63, 169–73, 179, 182–6, 188–92, 200–2.
  Data Doubles 47, 150
  Deeper 121–4, 130, 131–3, 135, 136, 201–2
  Forecasting 50, 150–1
  Mining 36, 40, 48, 52, 64, 85, 121–2, 126, 130, 147–50,

154, 156–7, 165, 175, 187, 189, 195
Insatiable 38
Power 41
Dataveillance 33–4, 41, 160, 182
Davos 112–13
Dialectic 42–5, 187–93
Digital
Civics 190
Orality 189
Tourism 95
Dyer-Witheford, Nick 34
Dystopia vii, 27, 34, 36, 39, 139, 142, 161, 196, 197

Electronic
Eye 32, 202
Electronic panopticon 45, 177, 202
Employment 16, 26, 36, 62, 73, 83, 86, 196, 199, 202
Employability 21, 81, 202
Entrepreneurial 21, 36, 45, 81, 89, 102–3, 113, 134, 146, 181–2, 185, 202

Facebook 1, 4, 8, 13, 37, 49, 51, 59, 61, 65, 72, 91, 134, 146, 165
Fantasy 17, 49, 77–82, 89, 106–8, 11, 134–5, 153, 158, 160, 183–4, 202
Fear of Missing Out 106, 145, 179
Financial Crisis xii, 82, 179
Foucault
Discipline xiii, xiv, 46, 48, 69, 71, 74–7, 109, 128, 156, 168, 172, 178–9
Heterotopia 104
Pastoral power 131
Social Technology 14, 24, 94
Foxconn 28
Future Attributes Screening Technology (FAST) 179

Ganesh, Shiv 9
Giddens, Anthony 43
Globalisation xi, 17, 22, 54, 65, 91, 100, 103, 112, 118, 142, 166
Goffman, Erving 12, 72
Front and back stage 'operating platforms' 72, 97
Great Recession 13, 21, 23, 117, 137

Hackers 9, 36, 98

Identity
Regulation 72
Smart 53, 63, 65–8, 79
Work 66, 72
Imperialism 31, 114
Intelligence
Artificial vii–viii, 42, 200
Business 151
Closed 166
Data 169
Inner 127–9
Internet
Addiction 8
Internet of things 71, 148
Intersectionality 10, 52, 55–6, 131

Klein, Naomi 113

Labour
Affective 132
Digital 4–5, 8
Lacan, Jacques
The Death Drive 133
Desire 184
Fantasy 78
History 158
Jouissance 111
Reality 106–7
Lefebvre, Henri 96
Luxemburg, Rosa 31

Marx (Marxists, Marxism)
Class struggle 16–18
Colonialism 103
Insatiable Capitalism 38, 88
Preference and Needs 180
Religion 115–16
Surplus Labour 111
The Matrix 133–4, 183
Mobile
Phones 69, 119, 165
Technology 13–14, 101, 104
Worlds 89–92
Munroe, Roland 98

Neoliberalism xiii, 6, 13–15, 17, 20–5, 40, 42, 44–5, 51, 68, 71–4, 102–5, 107–10, 113, 118–19, 140–3, 146, 175, 180

Obama, Barack 7
Ontological Security 43–5, 49 57–9, 63, 81, 92, 158

Panopticon 45–6, 49, 177
Post-Modernism 10, 13, 15, 55, 64–6, 68, 80, 96, 101–4, 154, 158
Power
Digital 36
Pastoral 131
Virtual 46, 50, 89, 104, 151, 161, 166, 179, 186–9, 195, 201–2
Prosumption 44, 173

Real-Time xiii, 11, 29, 48, 68, 70, 91, 95, 122, 134, 137, 144, 157, 161, 167, 179, 195
Tyranny 169–72
Ritzer, George
McDonaldization 44
Presumption 91
Rovelli, Carlo 140

Scott-Heron, Gil 186
Self/selves
Digital 11–14, 24
Multiple 6, 10–14, 41, 52, 55, 59, 62, 75–9, 80, 92, 99, 131, 148, 155, 166
Quantified 9, 30, 72, 87, 100, 104, 11, 154, 157
Saturated 58, 69, 130, 144, 159
Track 9, 20, 72, 122, 140, 154, 177, 182, 196, 200
Shackle, G.L.S. 139
Smart Technology
Governance 167, 170
Snowpiercer 141
Snowden, Edward 167, 187
Social media vii–x, 3–6, 11, 14, 20, 26, 40, 54, 56, 59, 62, 65, 72, 74, 77, 87, 90, 97–8, 113,

117–19, 124, 126, 128–30, 140, 142, 144, 148–50, 172–3, 177–9, 187, 192, 193–5, 199
Sousveillance 165–6, 188
Space
  Desakota 104
  Spatialisation 96–7
Spicer, Andre 86, 112, 126
  Optimisation 126
  The Wellness Syndrome 112
Spirituality 124–8
Surveillance
  Society 32, 167
  Surveillance-Industrial Complex 38
Synopticon 49
Technology
  Steam 32
  Wearable vii, 4, 29, 91, 131, 152
Thrift, Nigel 39, 115
Totalitarianism vii–viii, 162–74, 178–81, 183, 185, 187
Totalveillance 166, 172, 176, 178, 180, 182, 185, 187, 192
Trump, Donald 1–3, 117, 126
Tuckle, Sherry 66

World Economic Forum 31, 112

Zimmerman, Michael 79–80
Žižek, Slavoj 106
Zombi Politics 141
Zuboff, Shoshana 5